STRUCTURE AND BONDING

Volume 18

Editors: J. D. Dunitz, Zürich
P. Hemmerich, Konstanz · R. H. Holm, Cambridge
J. A. Ibers, Evanston · C. K. Jørgensen, Genève
J. B. Neilands, Berkeley · D. Reinen, Marburg
R. J. P. Williams, Oxford

With 43 Figures

Springer-Verlag
Berlin Heidelberg GmbH 1974

ISBN 978-3-662-15546-2 ISBN 978-3-540-37932-4 (eBook)
DOI 10.1007/978-3-540-37932-4

Originally published by Springer-Verlag Berlin Heidelberg New York in 1974
Softcover reprint of the hardcover 1st edition 1974

Library of Congress Catalog Card Number 67-11280.

Contents

STRUCTURE AND BONDING is issued at irregular intervals, according to the material received. With the acceptance for publication of a manuscript, copyright of all countries is vested exclusively in the publisher. Only papers not previously published elsewhere should be submitted. Likewise, the author guarantees against subsequent publication elsewhere. The text should be as clear and concise as possible, the manuscript written on one side of the paper only. Illustrations should be limited to those actually necessary.

Manuscripts will be accepted by the editors:

SPRINGER-VERLAG

D-6900 Heidelberg 1
P. O. Box 1780
Telephone (06221) 49101
Telex 04-61723

D-1000 Berlin 33
Heidelberger Platz 3
Telephone (030) 822001
Telex 01-83319

SPRINGER-VERLAG
NEW YORK INC.

175, Fifth Avenue
New York, N. Y. 10010
Telephone 673-2660

The Oxidation States and Reversible Redox Reactions of Metalloporphyrins

J.-H. Fuhrhop

Gesellschaft für Molekularbiologische Forschung mbH, 3301 Stöckheim über Braunschweig, and Institut für Organische Chemie der T. U. Braunschweig, 33 Braunschweig, Germany

Table of Contents

"The apartments were so irregularly disposed that the vision embraced but little more than one at a time. There was a sharp turn at every twenty or thirty yards, and at each turn a novel effect ... The second chamber was purple in its ornaments and tapestries, and here the panes were purple. The third was green throughout, and so were the casements."
E. A. Poe, The Masque of the Red Death

I. Introduction[1])

Redox reactions of metalloporphyrins usually can be followed visually: an electroneutral metalloporphyrin gives purple solutions, reductions or oxidation almost invariably change this color to green or brownish-green. The physical, chemical and biological applications of these reactions are very numerous and many natural scientists have had a part in the task of illuminating the purple and green chambers of nature as well as the many corridors connecting them from various viewpoints.

The classification of the redox reactions for metalloporphyrins includes four main types, according to the occurrence of:

1. Reversible changes in the formal oxidation number of the metal ("inorganic" redox reactions)

2. Reversible changes of the oxidation state of the porphyrin ("organic" reversible redox reactions)

3. Reversible or irreversible redox reactions of the central metal ion with axial ligands, in particular O_2.

---

[1]) Conventions used in this article: Nomenclature of porphyrins is explained in Section II_1. Redox potentials are always given versus aq. SCE; electronic spectra are given by the wavelength of the peaks in nm and their extinction coefficients, if available. Generally, only data which are not in *Falk*'s book (*56*) are given. Solvents and electrolytes used in spectroscopy and electrochemistry are mentioned only occasionally. Metalloporphyrin spectra are usually run in chloroform solution, oxidation potentials are obtained in butyronitrile, and reduction potentials in DMSO. Only the first author's name is given in the references in the text.

4. Reversible or irreversible redox reactions of the porphyrin ligand, in particular addition of O_2 or H_2.

The biological activities of metalloporphyrins can be classified similarly:

1. Electron transport is by virtue of a reversible valency change of the "inorganic" heme iron the main biological function of the cytochromes in various biological redox reactions (e.g. respiration, photosynthesis). The porphyrin ligand can always participate in the sequence of reactions.

2. The chlorophylls contain redox inert magnesium as the central ion and form "organic" π-cation radicals on oxidation. Such radicals play a prominent role in photosynthesis and the biosynthesis of protochlorophyll.

3. Some hemoproteins function in the transport and activation of molecular oxygen as well as in the decomposition of hydrogen peroxide (*e.g.* hemoglobin, tryptophan dioxygenase, cytochrome P450, catalases).

4. The metabolism of naturally occurring tetrapyrrole metal complexes includes irreversible oxygenation and hydrogenation reactions (bile pigment formation, protochlorophyll hydrogenation).

This review deals mainly with the chemical and physical aspects of the thermodynamically reversible redox reactions of porphyrins.

The relevant biochemical work will be referred to only briefly, although it should be kept in mind that many very detailed physical and chemical studies have been triggered by curious phenomena observed in biological systems containing hemoproteins or chlorophylls. It was the common interest of physicists, chemists, biologists, and physicians in the properties of metalloporphyrins, that led to the emergence of their redox chemistry as it is known today, and it is one of the purposes of this article to catalog the different approaches and techniques which have been used to illuminate these red and green chambers of the natural sciences within which so many scientific disciplines have some laboratory space. Early work on the redox chemistry of iron porphyrins has been covered admirably in at least three books [*Clark* (*33*), *Falk* (*56*), *Lemberg, Legge* (*121*)] and will be mentioned only briefly.

II. The Metal-Free Porphyrin Bases, their N-Protonated Dications and Deprotonated Anions

1. The Structure of the Porphyrin Ligands

The best known porphyrin ligand is certainly protoporphyrin, which in the form of one of its derivatives (Ia—d) is frequently used for redox experiments with metalloporphyrins.

Analogous properties are possessed by the synthetic porphyrins, like octaethylporphyrin (If, OEP), and etioporphyrin (Ig, ETIO); their properties help to avoid complications due to the reactive vinyl groups of PROTO and its derivatives or the formation of isomer mixtures in the case of reactions on these porphyrin macrocyclic ligands. All these ligands will be treated together as ETIO-type porphyrins in this review. The chemical properties of the unsubstituted porphin ligand (Ie, POR) are not known in detail. The methine bridges are also called mesopositions, and the peripheral carbons that bear the substituents are named β-pyrrolic positions.

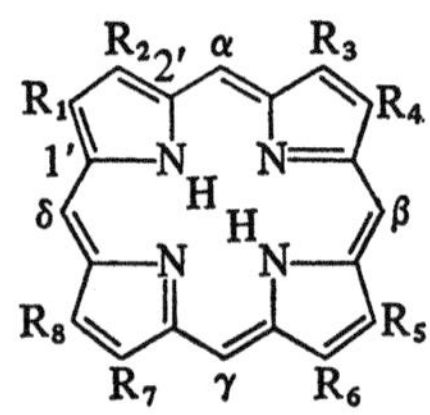

I

Table 1. Important porphyrins of the ETIO-type (β-pyrrolic alkyl substituents, no mesosubstituents)

Name of compound	Abbreviation	Substituents in the various β-pyrrolic positions							
		1	2	3	4	5	6	7	8
Protoporphyrin--dimethylester	Ia; PROTO	Me	V	Me	V	Me	P	P	Me
Deuteroporphyrin-dimethylester	Ib; DEUTERO	Me	H	Me	H	Me	P	P	Me
Mesoporphyrin--dimethylester	Ic; MESO	Me	Et	Me	Et	Me	P	P	Me
Haematoporphyrin-dimethylester	Id; HAEMATO	Me	CHOH | CH_3	Me	CHOH | CH_3	Me	P	P	Me
Porphin	Ie; POR	H	H	H	H	H	H	H	H
Octaethylporphyrin	If; OEP	Et	Et	Et	Et	Et	Et	Et	Et
Etioporphyrin I	Ig; ETIO*)	Me	Et	Me	Et	Me	Et	Me	Et

*) Only the ETIO I structure is given; the three other isomers presumably have identical redox chemistry.

P = propionic acid methylester
V = vinyl

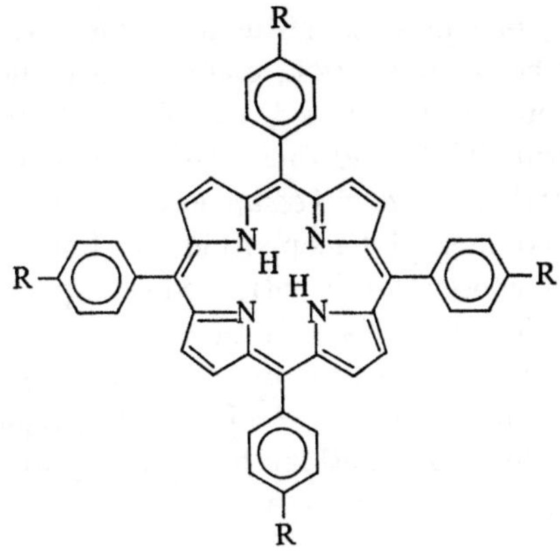

II

Another class of synthetic porphyrins comprises the meso-tetraphenylporphyrins, which can be easily prepared in large amounts, and which have some unusual properties, not found in natural porphyrins. Only the parent compound tetraphenylporphyrin (II R = H, TPP) with unsubstituted benzene rings is discussed here.

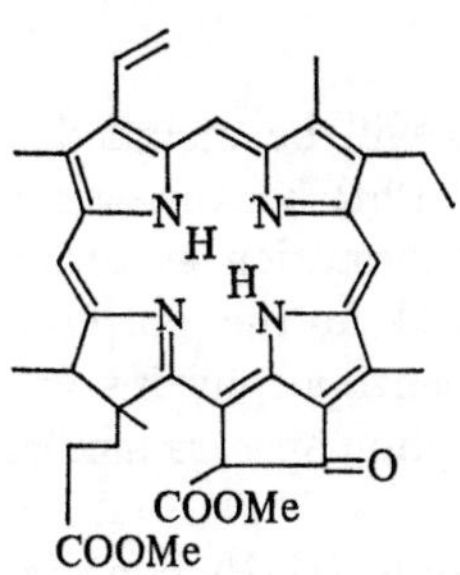

III

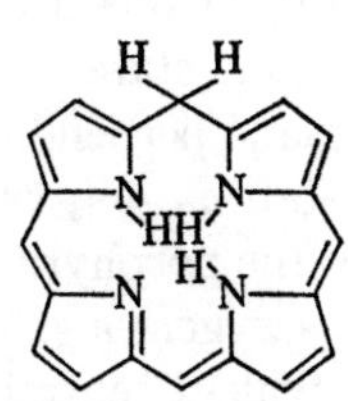

IV

The second major class of metalloporphyrins besides the hemes are the chlorophylls, with chlorophyll a (Chl_a) and bacteriochlorophyll (B Chl) as its main representatives. These compounds contain magnesium as the central ion, one meso-substituent in the form of the isocyclic ring and, as a further complication, one or two reduced pyrrole units. Porphyrins with one hydrogenated pyrrole unit are called chlorins (*e.g.* octaethylchlorin, OEC), and those with two opposite hydrogenated rings bacteriochlorins (*e.g.* Tetraphenylbacteriochlorin, TPBChl). Ligand III is called methylphaeophorbide a ($Pheo_a$). Its magnesium complex is usually chosen for in-vitro redox experiments and differs from the porphyrin ligand of Chl_a only by the esterifying alcohol on one proponic acid side chain (methanol instead of phytol). The phlorins (IV) constitute another important type of dihydroporphyrins. Here one hydrogen atom is added to a methine bridge of a porphyrin and another hydrogen atom is added to a nitrogen.

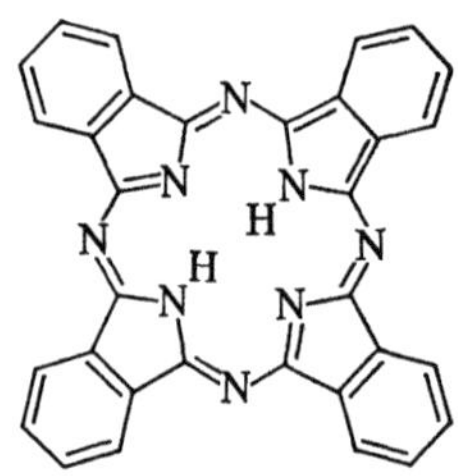

V

Finally some remarks will be included on the redox properties of metal phthalocyanines (V Pht) because a comparison with the metalloporphyrins yields some information as to how the different metal oxidation states are stabilized in the porphyrin cavity. Closely related porphyrins are the tetrabenzoporphyrins and the ms-tetraaza-porphyrins, which produce absorption spectra similar to those of phthalocyanines.

All of the above ligands (H_2P) contain on the central nitrogens two extremely weakly acidic protons, which can be substituted by two monovalent metal ions (M_2^{I}—P) or one divalent metal ion (M^{II}—P). If the central ion has an oxidation state of +3, an extra monovalent anion (X) is added to it in a plane perpendicular to the porphyrin plane ($M^{III}X$–P), if the oxidation state is +4, two extra monovalent anions join the metal on both sides of the porphyrin plane ($M^{IV}X_2$—P), or there is one divalent anion, usually oxygen, *e.g.* ($Mo^{IV}O$—P). Very often the coordination sites **5** and **6** of bivalent central ions are occupied by rather firmly bound neutral extra ligands. Familiar examples are CO, H_2O_2 and O_2 in iron, O_2 in cobalt, pyridine and water in almost all metal complexes of the porphyrins.

Some iron porphyrins have trivial names, defined for brevity as follows: hemes (or haems) are iron^{II} porphyrins (Fe^{II}—P); hemochromes (or hemochromogens) are iron^{II} porphyrins with nitrogenous bases as axial ligands (*e.g.* $Fe^{II}Py_2$—P); hematins (or haematins) are iron^{III} porphyrins with a hydroxyl group as axial ligand (Fe^{III}OH—P), whereas hemin (or haemin) contains a chloride anion (Fe^{III}Cl—P). Note that none of these designations includes a statement on the spin state of the iron ion.

X-ray analysis of many free porphyrin bases and their metal complexes showed that the overall geometry of the porphyrin macrocycle varies from almost planar in porphyrins with unsubstituted methine bridges (ETIO type) to "very ruffled" in the TPP derivatives [*Fleischer* (*65*)]. The porphyrins in a nonperturbing environment probably always exist in a near-planar conformation, but they are not completely rigid. Crystal interactions in the solid state as well as the effects of extra ligands in solution may cause severe deviations from planarity. This is true both of the free-base porphyrins, where protonations and deprotonations on the nitrogen atoms are the main sources of such disturbances, and of the metal complexes, where unsymmetrical additions of extra ligands to the central ion produce such effects. The bond distances in the macrocycle to the substituents of the porphyrin are rather invariant. A remarkable feature of its structure is the constant bond lengths (1.38 ± 0.01 Å) of the "inner" conjugation path containing the four nitrogens and the four methine bridges [*Hoard* (*94*)]. The lack of alternating bond lengths in this 16-membered ring is good evidence of its aromatic character (C bond length in benzene = 1.397 Å) and is one of the keys to the understanding of porphyrin physics and chemistry. Surprisingly, various central ions in the metalloporphyrins and the degree of protonation in the free-base porphyrins have little influence on these bond lengths [*Hoard* (*94*), *Fleischer* (*65*)]; the differences in the bond length for pyrrole rings with and without hydrogens never exceed ±0.02 Å (*Webb* 198a).

X-ray photoelectron spectra (ESCA) of TPP and some of its metal complexes have shown that the pyrrole nitrogens in the free base bear about 0.18 negative charge, whereas the pyrrolenine nitrogens are very close to neutral and the nitrogens in the metal complexes have an intermediate value [*Zeller* (*218a*)]. On the whole, these first ESCA results on porphyrins produce a confused impression because the spread of data is rather small and the differences found are often not interpretable [*e.g.* ZnTPP and $(Fe^{III}TPP)_2O$ have an identical high charge on the nitrogens (0.18), Fe^{II}(piperidine)$_2$ TPP has less (0.13) and Fe^{III} (Cl)TPP much less (0.04)].

Benzenoid hydrocarbons as well as porphyrins effect a considerable deshielding of their methine protons against an external magnetic field through molecular ring currents [*Jackman* (*105*)], as occurs also in the TPP and chlorophyll series. The electronic spectra of both classes of compounds are also comparable: they consist of very intense short-wavelength absorptions (the "Soret" band of porphyrins) and long-wavelength bands an order of magnitude less intense. Although the separation of the band is not analogous in both systems, the important features of the spectra can be described by the same simple circular box model [*Simpson* (*160*), *Malley* (*126*)]. TPP derivatives usually produce similar ETIO-type spectra, whereas in the chlorophylls and phthalocyanine derivatives both transitions occur with comparable intensity, indicating the lifting of the degeneracy of the outer (highest occupied and lowest empty) orbitals in these latter cases [*Gouterman* (*85*), *Weiss* (*199*)]. Many more analogies can be drawn between benzenoid hydrocarbons and porphyrins, relating to the chemical reactivity and the physical properties in the ground and excited states [*Fuhrhop* (*70*b)], but the examples given must suffice here.

2. The Ampholytic Character of the Porphyrin Ligands

A free porphyrin base is an ampholyte; it can either lose two protons in two consecutive steps or accept two protons [*Phillips* (*149*)]. The acidity of the two N—H protons is extremely weak. The two pk values have been estimated to be of the order of +16 [*McEwen* (*132a*)]. Titrations in organic solvents (*e.g.* dimethylsulfoxide or benzene with organic bases (*e.g.* tetramethylammoniumhydroxide or potassium tert. butylate in dimethylsulfoxide) do however, yield monoanions as well as dianions in high yield, provided that water is rigorously excluded [*Salek* (*155*)]. The protonated mono- and dications (*p*k ~2—4) form readily even in aqueous solutions (*149*). Dianions (VI) or dications (VII) contain two identical planes bisecting the nitrogen atoms, and this leads from the original C_{2v} symmetry to D_{4h} symmetry. D_{4h} is the symmetry of the porphyrin ligand in most of its metal complexes. The most obvious experimental result of this symmetry change is the collapse of the four-band visible spectrum of the porphyrin bases to a two-band spectrum between 520 and 600 nm in the metal complexes, dianions and dications of ETIO-type porphyrins [*Gouterman* (*85*), *Zerner* (*219, 221*)].

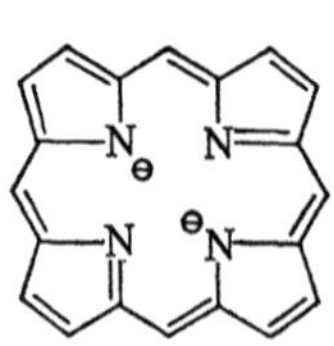

VI

VII

The ms-tetraphenyl derivatives behave differently: here the formation of a dication does not only lead to a reduction of the four-band neutral base spectrum, but also to a 50-nm bathochromic shift and a broadening of these bands, which are not found in any of the porphyrins with only partly substituted or nonsubstituted methine bridges. Even an ms-diphenyl porphyrin does not produce this effect [*Treibs* (*176*)]. The explanation is simple and straigthforward: in the neutral form the phenyl rings lie almost perpendicular to the porphyrin plane, whereas in the highly distorted structure of the dication (arising from van der Waals and Coulomb repulsions) they rotate towards a more parallel position, which allows the overlap of π orbitals from the porphyrin and phenyl rings. This has also been shown by X-ray analysis of the dications [*Fleischer* (*65*)]. Already the monocation is probably far from planarity, as is indicated by the immediate further protonation of this species. This extreme flexibility of the porphyrin plane has so far been demonstrated for TPP derivatives but not for ETIO-type porphyrins.

The monocations and anions of porphyrins produce complicated electronic spectra [*Falk* (*56*), *Salek* (*155*)], which indicate a less symmetric chromophor than in the neutral base or the dications and dianions. One might therefore conclude that the "uneven" proton is localized at one nitrogen atom and not sitting in the center of the porphyrin cavity like a metal ion in the metalloporphyrins.

3. Electronic Properties of the Porphyrin Ligands

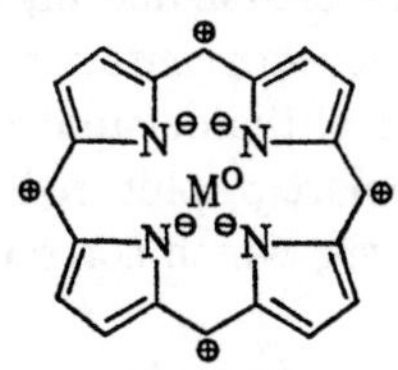

VIII

Woodward has pointed out that porphyrins contain pyrrole units that are on average one half of an electron away from a stable 6-π-electron configuration [*Woodward* (*214*)]. From the presumed tendency to become aromatic pyrrole units one can deduce a metalloporphyrin resonance structure like (VIII), where the methine bridge carbons have lost π-

electron density in favor of the pyrrole rings. The rather extreme formulation (VIII) is useful because it helps to visualize the most important experimental finding on ETIO-type metalloporphyrins. This is the stabilization on the central metal ion of high positive charges, which are neutralized by the negative nitrogens. Furthermore, it is a simple model for the rationalization of the behavior of related conjugation systems: chlorins and bacteriochlorins contain reduced pyrrole units with no tendency to pull electrons from the methine bridges and therefore should have less negative charge on the nitrogens, whereas in the phthalocyanines the electronegative bridge nitrogens should impede with polarizations as indicated in (VIII). TPP derivatives are predicted to be of varying basicity, depending on the positioning of the phenyl substituents relative to the porphyrin plane. If, therefore, the major metal–porphyrin interaction consists of electron donation from the ligand to the metal (and this will later be shown to be the case) then an approximate order ETIO > TPP > Chl_a > Pht can be established. This should represent the relative ability of the ligands to stabilize higher oxidation states of metal ions versus their lower oxidation states. *Gouterman's* early MO calculations on various substituted porphyrins yielded qualitatively the same order of basicity (*85*). *Pullman* pointed out that in "covalent" metalloporphyrins the lowest empty π molecular orbital is of high energy, so that the π system does not have any marked electron acceptor properties, or in other words, is a weak π-acid. (*149a*).

4. The Redox Behavior of Free Porphyrin Bases, their Protonated Dications and Deprotonated Dianions

Usually the chemical oxidation of porphyrins and their deprotonated anions and protonated cations is not reversible. Poor reductants and oxidants do not induce any reaction, whereas the action of powerful reagents usually leads to irreversible decomposition. There are, however, some reports on unresolved, low-percentage radical esr signals in solutions of free-base porphyrins and phthalocyanines [*George* (*81*), *Gibson* (*82*)], and there are reports of polarographic redox potentials of some non-metallic porphyrin derivatives, which correspond to reversible reactions (Tables 2 and 3).

The general rule that can be extracted from these tables is that porphyrins of higher electron density are easier to oxidize and more difficult to reduce than the more electropositive porphyrins. This charge effect is most pronounced in the photochemistry of the cations [*Umrikhina* (*187*), *Mauzerall* (*130*)] and anions [*Fischer* (*63*)]. Introducing substituents which disturb the planarity of the porphyrin ring tends to facilitate reduction and impede oxidation.

Table 2. Oxidation potentials of some nonmetallic porphyrin derivatives

Porphyrin	$E^{ox}_{\frac{1}{2}}(1)$	$E^{ox}_{\frac{1}{2}}(2)$	Literature	
OEP	0.81	1.30	(*74*)	*Fuhrhop*
ETIO	0.77	—	(*166, 167*)	*Stanienda*
PROTO	0.83	—	(*166, 167*)	*Stanienda*
TPP	0.97	(1.12)	(*166, 167*)	*Stanienda*
α-Oxo-OEP	−0.05	—	(*73*)	*Fuhrhop*
$OEPH_2^{2+}$	1.65	—	(*74*)	*Fuhrhop*
N-Methyl-OEP	1.37	—	(*74*)	*Fuhrhop*
N-Methyl-OEP anion	0.86	—	(*74*)	*Fuhrhop*
$Pheo_a$	0.86	1.17	(*166*)	*Stanienda*

Note: The hydrogen atoms on the nitrogens of the neutral porphyrin bases have been omitted.

Table 3. Reduction potentials of some nonmetallic porphyrin derivatives

Porphyrin	$E^{red}_{\frac{1}{2}}(1)$	$E^{red}_{\frac{1}{2}}(2)$	Literature
OEP	−1.46	−1.86	(*74*) *Fuhrhop*
ETIO	−1.37	−1.80	(*32, 60*) *Clack, Felton*
MESO	−1.34	−1.73	(*147*) *Peychal-Heiling*
TPP	−1.08	−1.52	(*32, 60*) *Clack, Felton*
$TPP^{2-}\ 2Na^{+}$	−1.7	—	(*60*) *Felton*

III. Some General Remarks on the Oxidation States of Metalloporphyrins (Electronic Spectra, esr Spectra, Electrochemical Regularities)

The description and discussion of the oxidation states of metalloporphyrins is complicated by the fact that not only the central metal ion but also the π-conjugated system of the porphyrin ligand is redox-active. However, in almost all cases investigated so far it has been possible to assign either a metal or a ligand reaction to a given oxidation or reduction step. In perhaps two or three of the 85 metalloporphyrin oxidation states reported so far evidence has been produced that an electron that had been removed (or introduced) was common to both the metal and the porphyrin. For example, if a one-electron oxidation of an electroneutral metalloporphyrin (M^{II}—P) occurs, either a porphyrin π-cation radical is formed or the central ion is raised to the trivalent state. The possible hybrid oxidation states found so far are the Mn^{III}, Mo^{V} and Ni^{III} porphyrins. Evidence for these statements will be given in all the singular cases of redox reactions. The reason for the generally simple redox behavior of metalloporphyrins is, of course, that the metal $d(xz,yz)$ orbitals and porphyrin π orbitals do not mix to any great extent.

In this section we give the analytical tools and arguments that help us to decide whether it is the porphyrin ligand or the metal ion that has undergone change after a given redox reaction.

From the slope of a potentiometric titration curve or of a polarographic step and, in favorable cases, from elemental analysis of the products, we know whether we are dealing with a one- or two-electron exchange. With very few exceptions metalloporphyrins are oxidized or reduced in well-defined, one-electron steps separated by more than 100 mV.

Electronic Absorption Spectra

The next important information on the product is derived from the electronic absorption spectrum. This can usually be obtained by electrolysis of a dilute solution of the metalloporphyrin at a constant potential or by oxidometric titrations directly in the absorption cell. Most redox reactions with metalloporphyrins give good isosbestic points when absorption spectra are taken at various stages of oxidation, and they are fully reversible when no chemical addition reaction to the porphyrin ligand has occurred. Ten typical absorption spectra of metalloporphyrins are given below and correlated with the various metalloporphyrin oxidation states.

1. The "Normal" Metalloporphyrin Spectrum. When this type of spectrum with sharp intense Soret and visible bands is found for an oxidation or reduction product from a metalloporphyrin, we can be certain that the porphyrin ligand is not touched and only the metal has changed its oxidation state (*e.g.* $Co^{II} \rightleftharpoons Co^{III}$). [*Falk* (*56*)]. An exception is the case of π dianions in aprotic solvents, which show a similar, although broadened spectrum [*e.g. Closs* (*35*)]. Fe^{II} porphyrins (heme, hemochromes) are also considered to be of the normal spectral

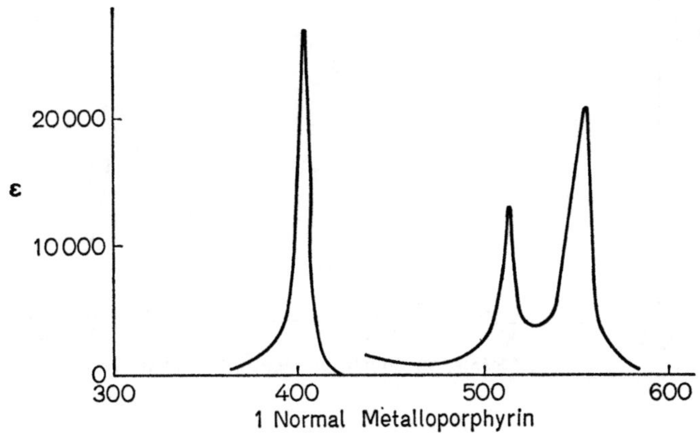

Fig. 1.1

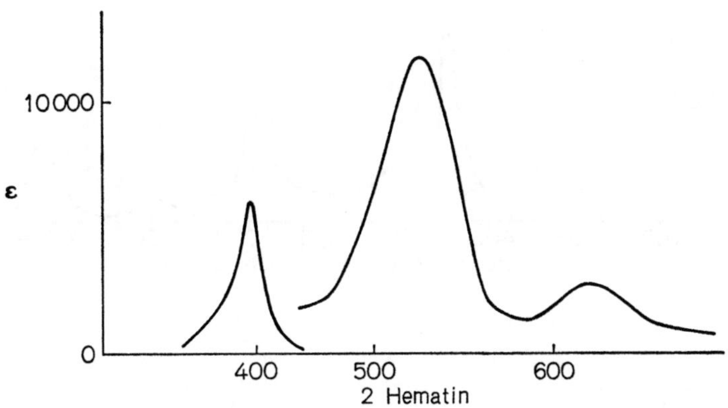

Fig. 1.2

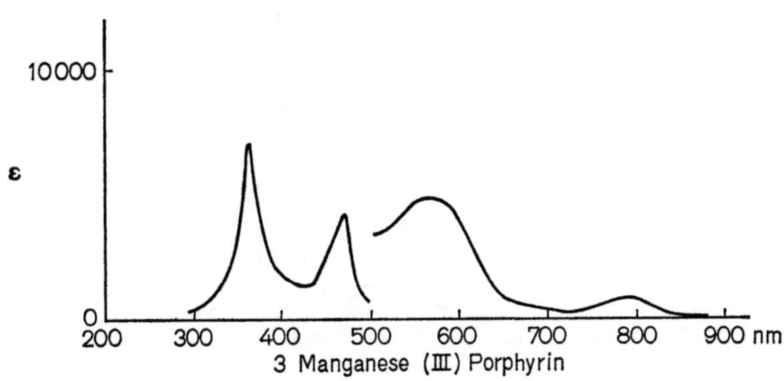

Fig. 1.3

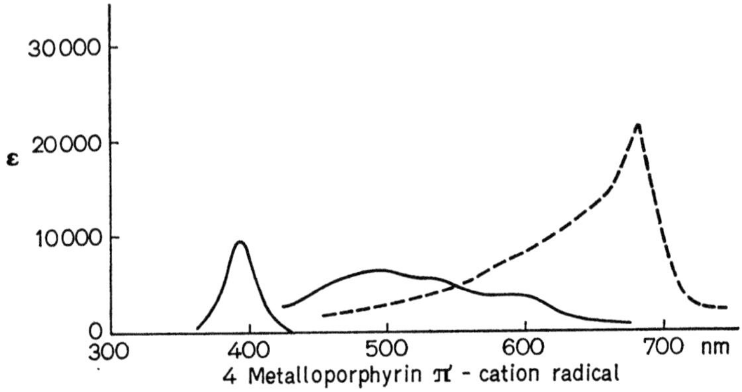

Fig. 1.4

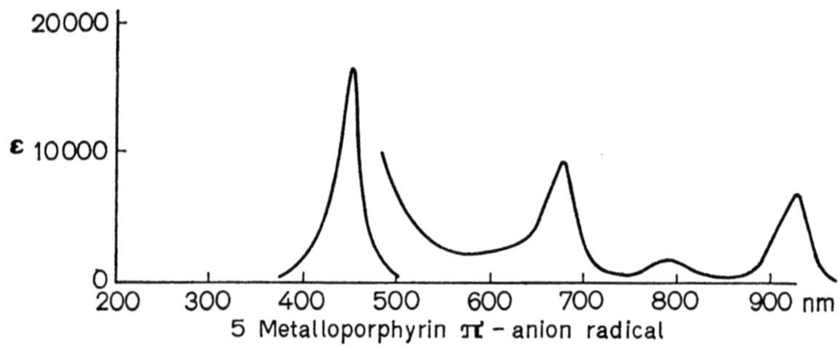

Fig. 1.5

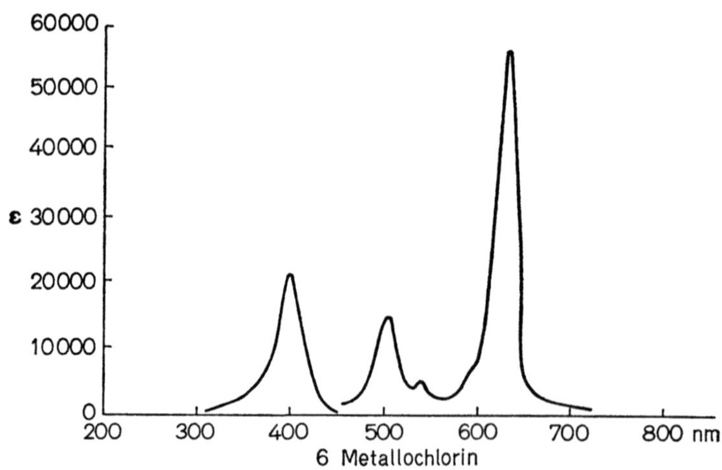

Fig. 1.6

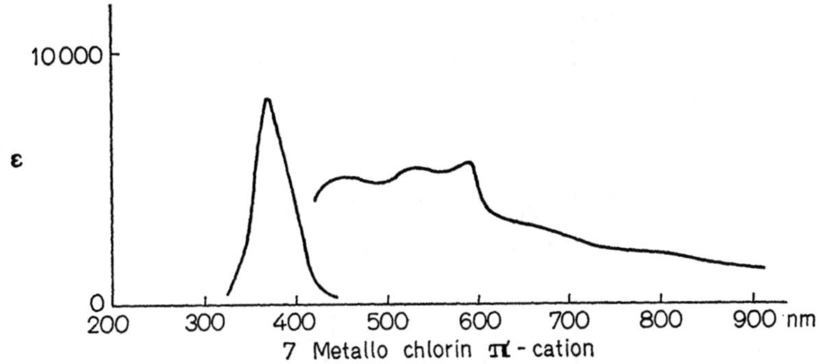

Fig. 1.7

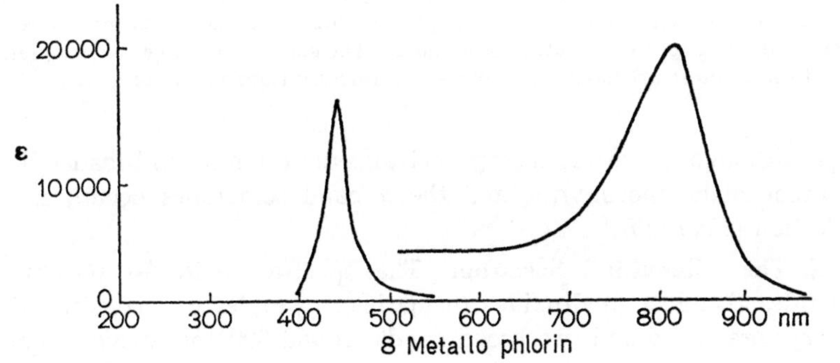

Fig. 1.8

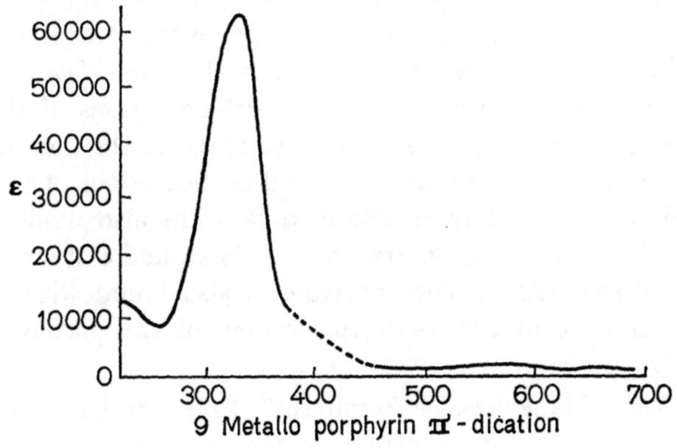

Fig. 1.9

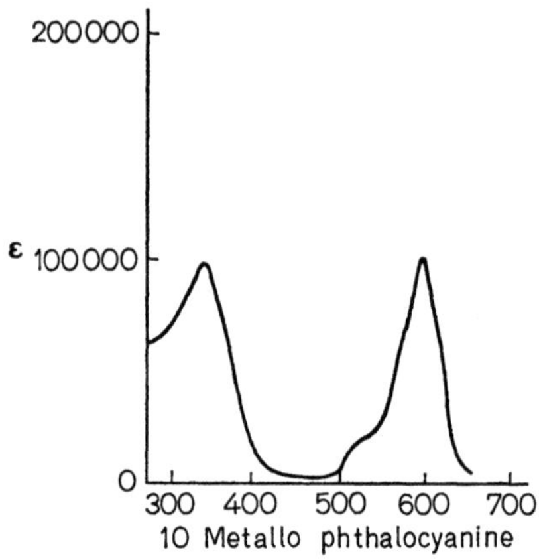

Fig. 1.10

Fig. 1.1—1.10. Typical spectra of metalloporphyrins in various oxidation states. The extinction coefficients on the left of the spectra (Soret band) have to be multiplied by a factor of 10 except for spectra nos 9 and 10. The spectra are to be taken as semi-quantitative illustrations to the more general descriptions in the text

type, although the visible absorption bands are often much broader than in other metalloporphyrins and the α band sometimes occurs as an inflection [*Falk* (*56*)]^{I}.

2. The "Hematin" Spectrum. The spectra of iron^{III} porphyrin chlorides (hemins), hydroxides (hematins), cyanides etc. consists of a strong Soret band and two broad bands around 500 and above 600 nm, respectively. Such bands also occur with certain porphyrin complexes of three- to five-valent transition metals containing axial ligands unsymmetrical with respect to the porphyrin plane. The perturbation of the "normal" spectrum is probably caused by intramolecular charge transfer from the porphyrin π orbitals to the metal $d_{xz,yz}$ orbitals [*Zerner* (*221*)]. A change in the spin state of the central ion (*e.g.* in high- and low-spin iron complexes) usually does not lead to pronounced differences in the spectrum of the π-π^*-transitions of the metalloporphyrins. However, changes are always found in the fine structure of the electronic spectra and have been used as indications of the spin state in individual cases where magnetic data were also available [*Williams* (*205*)]. Weak extra charge-transfer bands around 800 nm are sometimes found [*Zerner* (*221*)]. The spectrum is also found with an unchanged oxidation state of the π-electron system of the porphyrin ligand *e.g.* [*Falk* (*56*)].

3. The "Manganese^{III} Porphyrin" Type Spectrum. The main features of this type of spectrum are the split Soret band, broad visible absorptions similar to the hematin type, and strong charge-transfer bands around 800 nm. Again this indicates a porphyrin π-electron system in its zero-oxidation state [*Boucher* (*11*)]. A Soret band at ~470 nm is also found when the ionic radius of the central ion exceeds 1 Å (*e.g.* Pb^{II}).

4. The metalloporphyrin "π-cation radical" spectrum. An extremly broad band covering the whole visible range of the spectrum and often extending into the near infrared is typical of a one-electron oxidation product of the porphyrin ligand. Sometimes strong visible absorption bands around 700 nm are also found with these products, but the broad visible absorptions are always present. The Soret band usually has a low extinction coefficient and is broadened considerably [*e.g. Fuhrhop* (*76*)].

5. The Metalloporphyrin "π-Anion Radical" Spectrum. The one-electron reduction products of the porphyrin ligand produce a strong absorption above 800 nm and a broad, structured Soret band [*e.g. Closs* (*35*)].

6. The "Metallochlorin" Spectrum and its "π Anions". The spectrum of the chlorins and bacteriochlorins, where one pyrrole ring or two opposite rings of the porphyrin ring are hydrogenated, consists of four visible bands with a very intense α-band and a sharp Soret band of relatively low intensity. The "π anions" produce a similar spectrum with bathochromic shifts of the long-wavelength band of about 100 nm [*Fajer* (*55a*)].

7. The "Metallochlorin π-Cation" Spectrum. This is very similar in appearance to the metalloporphyrin π-cations spectrum although there is usually even less structure in the visible absorption [*e.g. Fuhrhop* (*70a*)].

8. The "Metallophlorin" Spectrum. This is the spectrum of a metalloporphyrin in any oxidation state, which has added an extra substituent at one methine bridge. It is also found very often in various oxidation or protonation products of tetraphenylporphyrins. It is very similar to the spectrum of metallo-bilatrienes with an opened macrocyclic ring [*Wasser* (*195*) *e.g. Peychal-Heyling* (*147*)].

9. The Metalloporphyrin π-Dication Spectrum. This is usually very similar to π-cation radical spectra, but the Soret band has shifted below 350 nm and is even less intense. Often phlorin formation is observed [*e.g. Fajer* (*55*)].

10. The Phthalocyanine Spectrum. Although the conjugation pathways of phthalocyanines and porphyrins are comparable, as is the strength and geometry of the ligand field, the spectrum is more similar to that of polyene-type chromophors with no degeneracy in the outer orbitals. Ms-tetra-azoporphyrins and tetrabenzoporphyrins also produce spectra with high- and low-energy absorption bands of comparable intensity [*e.g. Lever* (*122*)].

A word of caution is in place here: only the spectral types 1,4,6,8 and 10 have been described in sufficient different compounds for their descriptive power to be relatively certain. The other spectra have been produced in a very limited number of cases, so that different spectra could emerge for chemically closely related species.

Indeed, the correlation of the electronic spectra with different metalloporphyrin states is not unequivocal in many specific cases. The distinctions between 2 and 4 are not clearcut, as is the case with 5, 8 and 9. Further complications arise when the metalloporphyrin aggregates, which usually leads to absorption-band broadening. In many of these cases magnetic data will provide additional evidence, or certain electrochemical regularities (see V.1) can be used.

The usually strong belief of porphyrin chemists in the analytical usefulness of electronic spectra for positive identification is by no means justified for the many new species with oxidized or reduced π-electron systems characterized during the last five or six years.

Esr Spectra

In a paramagnetic metalloporphyrin the unpaired electron may be localized in the porphyrin-ligand π orbitals, or the metal d orbitals, or in both. Accordingly, the esr experiment with these metal complexes can yield three different results: 1. an isotropic signal characteristic of an organic free radical around $g = 2.00$ and of linewidth less than 12 G; 2. an anisotropic signal of linewidth $\geq$ 50 G with various g values typical for transition metal ions; 3. no signal. So far, no triplet-state spectrum has been reported for species with a paramagnetic central ion and a porphyrin radical ligand. Each case will now be briefly discussed.

1. Porphyrin Radicals with Diamagnetic Central Ions. Two types of hyperfine splitting patterns are observed in porphyrin cation-radical esr spectra: the ETIO-type cation radical, and the TPP-type cation radical. Two distinct differences are found between them: 1. the large linewidth (~5G) in the ETIO-type cation radicals as compared to the narrow linewidth (100—200 mG) in the TPP radicals; 2. the fact that in the ETIO radicals only methine proton couplings ($MgOEP^{\dot{+}}$: 1.48 G are observed and no N^{14} nitrogen couplings, whereas TPP cation radicals do produce nitrogen hyperfine couplings ($ZnTPP^{\dot{+}}$: 1.58 G). This difference has been interpreted by the simple and straightforward HMO argument, that in one case the a_{1u} orbital with no electron density at the nitrogens contains the unpaired electron in the ground state (*e.g.* $MgOEP^{\dot{+}}$) whereas in the other case the unpaired electron is in the a_{2u} orbital [*Borg* (*10*), *Dolphin* (*49*), *Fajer* (*55*)]. The couplings of the

Fig. 2. Esr spectrum of an ETIO-type magnesium porphyrin cation radical. The hyperfine couplings are from the methine protons

methine bridge protons are, however, larger than those expected from these arguments. In the ETIO porphyrins it is likely that the unpaired electron resides in an orbital which can be derived from a linear combination of nearly degenerate a_{1u} and a_{2u} levels (thermal averaging). The hyperfine couplings and the large line-width can be explained better by a thermally averaged MO for the unpaired electron than by individual a_{1u} or a_{2u} orbitals. Substitution of four phenyl rings in the bridge position raises the level to a_{2u}, thus lifting the degeneracy. One can then explain the presence of the unpaired electron in this orbital as described above. Much more detailed experimental and theoretical work is needed to prove such connections between the relative energies of the outer orbitals and the nature of different central ions, meso-substituents, solvents etc., and to interpret the meaning of these relationships.

The anion radicals of metalloporphyrins are reported to give rise to broad single-line esr signals. They are thought to arise from the presence of an unpaired electron in the doubly degenerate e_g orbital [*Hush* (*97*, *98*)]. Phlorin anion radicals yield esr spectra with extensive hyperfine splittings which have not so far been interpreted [*Hush* (*97*)].

2. Metalloporphyrins with one Unpaired Electron on the Central Metal Ion. EPR signals, which are due to the presence of paramagnetic transition metal ions in the porphyrin cavity, exhibit strong anisotropy in the g values and the hyperfine couplings, always typical of a transition metal ion in a square planar ligand field. Superhyperfine couplings from the porphyrin nitrogens are also observed in favorable cases. Since for the transition metal ions the spin—orbit interactions are the main source of spin relaxation, linewidths in these cases are much larger than those obtained in the case of free radicals. (Fig. 3)

3. Metalloporphyrins with Two or More Unpaired Electrons. The frequent case where no esr signal is detectable in a paramagnetic metalloporphyrin has also to be discussed here. This occurs when both the metal and the porphyrin ligand contain unpaired electrons. The resulting

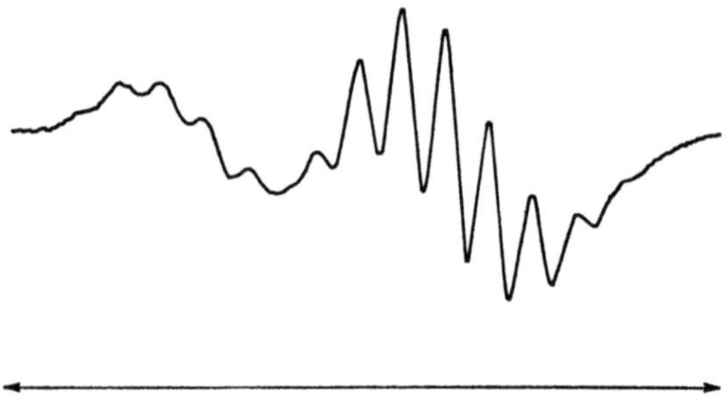

Fig. 3. High-field half of a typical metalloporphyrin esr spectrum with a paramagnetic central ion (Cu OEP)

triplet or higher state usually does not produce any esr absorptions down to the temperature of liquid nitrogen (*e.g.* Cu $OEP^{\dot{+}}$) [*Fuhrhop* (*76*), *Wolberg* (*212*)]. Metal ions in doublet or multiplet spin states with very short electron relaxation times also do not give detectable esr signals. Aggregation of paramagnetic metalloporphyrins often precludes the observation of an esr signal, because of extensive line broadening (*e.g.* Mn^{III} (OH)OEP) due to spin–spin interactions [*Boucher* (*11*), *Yonetani* (*217*)].

In such cases measurements of the bulk suspectibilities often yield valuable information on the overall oxidation state of the metalloporphyrin as compared to the starting material [*Havemann* (*90*), *Fuhrhop* (*71*), *Wolberg* (*212*)].

Electrochemical Regularities

The important rules are that the normal difference between the first one-electron oxidation and reduction potentials of the porphyrin ligand is 2.25 ± 0.15 and that the second removal or addition of an electron follows after an interval of approx 0.3 V [*Fuhrhop* (*74*), *Kadish* (*109*)] (see V.1).

IV. The Oxidation States of the Central Metal Ions in Metalloporphyrins

1. Group-III a Porphyrins

Scandium

Sc^{III}. Only the $Scandium^{III}$ complex of OEP has been reported [*Buchler* (*19*)]. It is diamagnetic and produces a normal electronic spectrum [Sc OEP(OAc), **574** (**16,600**), **538** (**12,300**), **406** (**332,000**)]. This complex is interesting because it is the only metalloporphyrin, with a higher oxidation state than +2 on the central ion perhaps: which is known to decompose with acetic acid. Similar properties are found in the rare-earth phthalocyanines [*Lever* (*122*)]. This inherent lability together with the large radii of these ions may explain why it has not so far been possible to prepare the rare-earth metalloporphyrins.

Moreover, a scandium porphyrin is the only metalloporphyrin for which theoretical predictions of its properties were available in advance of the actual isolation of the product [*Zerner* (*220*)]. The results of these calculations were unusual: they predicted an electronic charge transfer in the excited state from the porphyrin ligand to the scandium metal, as well as a strong interaction between the porphyrin π orbitals and the metal $d(xz,yz)$ orbitals. An internal electron transfer should even induce photolability in this compound. On the other hand, it was predicted that the covalent bonding in the ground state between the porphyrin nitrogens and scandium should be as strong or stronger than that calculated for the vanadyl complex. None of these results survived experimental checking, and the failure of the extended Hückel method with this simple transition metalloporphyrin (*a* d^0 state) may be taken as a warning. The large amount of experimental material that has been gathered in the field of metalloporphyrins gives quite a detailed picture, whereas nowadays MO calculations deal, in the words of the authors [*Zerner* (*219*)], with "shadows on the wall". Only if all known experimental facts are built into the fundamental assumptions of the MO model, might one expect the shadows to acquire a meaningful structure and the detailed predictions to become realistic.

2. Group-IVa Porphyrins

Titanium

Ti^{IV}. Diamagnetic Ti porphyrins have been isolated and characterized as titanyl (Ti=0) complexes [*Fuhrhop* (*71*), *Tsutsui* (*181*), (*183*)]. An esr spectrum of a Ti^{III} state has been mentioned, but no detailed data are available on this oxidation state nor on the potential of the Ti^{III}/Ti^{IV} couple [*Tsutsui* (*183*)]. The "normal" electronic spectrum of titanyl OEP [573 (28,000), 535 (14,000) 405 (350,000)] [*Fuhrhop* (*71*)] is replaced by a "hematin" spectrum [582 (8,500) 480 (24,000) 359 (105,000)] [*Fuhrhop* (*71*)] and the fluorescence of the original titanyl compound disappears when the axial oxo group is protonated. The influence on the redox potential of the intramolecular $\pi \rightarrow d_{xz,yz}$ charge transfer, which occurs similarly in the hematins (see IV.6), has not been investigated.

Zirconium, Hafnium

Zr^{IV}, Hf^{IV}. Only diamagnetic metalloporphyrins with "normal" electronic spectra and oxidation states +4 with two axial ligands have been reported. [Hf^{IV} OEP (0 Ac)$_2$: 569 (42,500) 530 (25,000) 401 (420,000) Zr^{IV} OEP(OAc)$_2$: 567 (32,000) 529 (12,000) 400 (380,000)] [*Buchler* (*19*, *Buchler* (*21*)].

3. Group-V a Porphyrins

Vanadium

V^{IV}. As with the titanium porphyrins, these complexes occur as vanadyl complexes with an axial oxo group and produce a "normal" electronic spectrum which changes to a "hematin" spectrum when trifluoroacetic acid is added [*Bonnett* (*9*)]. The vanadyl porphyrins contain a single unpaired electron ($\mu = 1{,}4$ B.M.) and their esr spectra have been studied in some detail [*Kivelson* (*113a*), *Roberts* (*152*)]. The rich redox chemistry of the inorganic vanadium salts is not reflected in the vanadyl porphyrins, where the +4 state seems to be exceptionally stable.

Niobium, Tantalum

Nb^{V}, Ta^{V}. Both diamagnetic metalloporphyrins have been synthesized with three monovalent ligands in the +5 state and have "normal" electronic spectra [Nb OEP F_3: 569 (24,000) 532 (8,300) 403 (186,000); Ta OEP F_3: 576 (27,000) 528 (8,300) 399 (174,00)] [*Buchler* (*21*)].

4. Group-VI a Porphyrins

Chromium

Cr^{II}. Porphyrins have been prepared in solid form and show a "normal" spectrum in solution [Cr^{II} MESO: 572 (4,400) 516 (1,350) 415 (47,000)] [*Tsutsui* (*183*)]. The magnetic susceptibility of the solid was found to be 2.84 BM, in solution 4.95 BM. The latter value corresponds to four unpaired electrons; the low susceptibility in the solid state has been explained by intermolecular interactions.

Cr^{III}. Porphyrins are easily formed by air oxidation from the Cr^{II} compounds [*Tsutsui* (*183*)], or directly from porphyrins and Cr^{III} acetylacetonates [*Buchler* (*19*)], and show a "normal" spectrum with broadened bands in the visible but a complicated Soret band which somewhat resembles the Mn^{III} spectrum [Cr^{III} (Br) OEP: 576 (4,300) 546 (9,800) 457 (43,000) 444 (45,000) 370 (51,000) 354 (47,000)] [*Buchler* (*19*)].

Cr^{IV}. Porphyrins have not yet been obtained in pure form either in solution or in solid form, but their existence is thought probable because in voltammetric measurements an extra oxidation step is found at 1.0 V [*Kadish* (*109*)].

Molybdenum

Mo^{V}. The first reported molybdenum porphyrins were the Mo tetraphenylporphyrins obtained from $Mo(CO)_5$ and TPP in the presence of air [*Fleischer* (*67, 68*)]. The simpler octaethylporphyrin complexes have also been described [*Buchler* (*19*)]. Molybdenum^{V} porphyrins contain an axial oxo group and a monovalent ligand; their magnetic susceptibility points to one unpaired electron (MoO(Cl)TPP = 1.74 BM), and esr spectrum corresponding to about one unpaired electron per molecule with hyperfine splitting through the nitrogens is found in solution [*Fleischer* (*68*)]. The spectrum is of the "Mn^{III}-porphyrin" type with a split Soret band. (MoO(OH)OEP: 595 (9,600) 562(14,500) 443(83,000) 342(49,000)] [*Buchler* (*19*)].

Mo^{IV}. Sublimation of Mo^{V} porphyrins [*Buchler* (*19*)] or their reduction with H_2S leads to diamagnetic Mo^{IV} complexes with a normal electronic spectrum [578 (21,000) 540 (12,000) 412 (182,000)] [*Buchler* (*19*)]. The diamagnetism of this d^2 system clearly points to a d_{xy} orbital which is much lower in energy than the $d_{xz,yz}$ orbitals, so that the two electrons are paired in the nondegenerate orbital. The same is true for the VO porphyrin system, where the unpaired electron is also in the d_{xy} orbital [*Kivelson* (*113a*)], and the molybdyl phthalocyanines [*Hill* (*92a*)]. A qualitative argument to explain this somewhat unexpected result is that the metal–oxygen "double bond" produces a high electron density along the z axis and thus raises the $d_{xz,yz}$-orbital energies sharply over the d_{xy} orbital. The double bond in titanyl, vanadyl and molybdyl porphyrins is also exemplified by the presence of a strong absorption band at approx. 965 cm^{-1}.

5. Group-VIIa Porphyrins

Manganese

Only with iron has the redox chemistry in the porphyrin series been worked out in more detail than with manganese. It is mainly the groups of *M. Calvin* and *L. J. Boucher* who have investigated these metalloporphyrins, and extensive reviews are available from both [*Boucher* (*11*), *Calvin* (*22*)]. The interest of these complexes lies in the fact that manganese is always found in photosynthetically active plant material and seems to be active in the oxidation of water [*Park* (*138*)]. Because of the ubiquity of metalloporphyrins in biological redox systems, it was thought that a good first model of these complexes would be the manganese porphyrins [*Loach* (*124*)]. Although this hypothesis has not been substantiated since it was first proposed in 1963, it triggered some useful work on metalloporphyrins with exceptional properties, which might otherwise not have been undertaken.

Mn^{II}

Mn^{II} porphyrins are obtained in solution by reduction of Mn^{III} complexes with sodium dithionite in the presence of pyridine [*Calvin* (*23*), *Loach* (*124*)], or in the solid state as a thin layer by evaporation of Mn^{III} porphyrins in high vacuum [*Calvin* (*22*), *Edwards* (*52*)]. No preparation of crystalline materials has been reported so far.

In solution a magnetic susceptibility of 5.8—5.9 BM is found, which corresponds to a d^5 high-spin state [*Loach* (*124*)]. Molecular-orbital calculations indicate that a planar configuration will give rise to an intermediate spin state and not a d^5 state for Mn^{II}. The fact that the antibonding metal $d_{x^2-y^2}$ orbital is singly occupied has been taken as evidence for Mn^{II} being out of the plane [*Boucher* (*11*)]. The electronic spectrum corresponds to a normal spectrum of a metalloporphyrin with two sharp bands in the visible and an intense Soret band [*Loach* (*124*)]. The acid lability of the Mn^{II} porphyrins is comparable to that of the magnesium complexes. They are demetalated by acetic acid. The addition of pyridine to Mn^{II} porphyrins in the solid state leads to a 40 nm bathochromic shift of the Soret band and a rise in intensity [*Yamamoto* (*216*)]. The stoichiometry of this reaction has not been investigated, but the dramatic long-wavelength shifts suggests, that only one pyridine is added to the central manganese ion, pulling it further out of the porphyrin plane. The X band esr spectrum of Mn^{II} protoporphyrin, embedded in proteins or Sephadex and measured at liquid nitrogen temperature, revealed two signals at $g=2.0$ and $g=5.9$ as well as a zero field splitting constant $D=0.5$ cm^{-1} [*Yonetani* (*217*)].

Mn^{III} and the Redox Couple Mn^{II}/Mn^{III}

The most stable manganese oxidation state in porphyrins is +3, and Mn^{III} complexes of ETIO-type porphyrins are the most prominent example of metalloporphyrins with a split Soret band. Instead of the usual intense band at 400 nm, these chelates possess two bands at 350 and 460 nm with an intensity ratio of 2:1. This change was related to a strong porphyrin–metal π–$d_{(xz,yz)}$ interaction in addition to the usual σ interaction between the four pyrrole nitrogens and the metal [*Boucher* (*11*)]. Theoretical calculations do indeed predict $d_{xz,yz}$–π interactions because the metal d orbitals in the first transition series roughly match the porphyrin levels [*Zerner* (*219*)], but the dramatic singularity of the Mn^{3+} in this series is not understood. The relatively large zero field splitting of Mn^{III} PROTO ($D=7.6$ cm^{-1}) has been determined by infrared measurements [*Richards* (*150a*)], and no esr absorption at liquid nitrogen temperature could be observed in the X band, probably for this reason.

Electrochemical reductions of the central Mn^{III} ion have been performed with many different porphyrin and axial ligands. Some selected values are given in Table 4. The order of the $E^{red}_{1/2}$ for different macrocyclic ligands is Phth < ETIO < TPP < Pheo a.

Table 4. Reduction potentials of some Mn^{III} porphyrins

Compound	$E^{red}_{\frac{1}{2}(1)}$ ($Mn^{III} \rightleftharpoons Mn^{II}$)	$E^{red}_{\frac{1}{2}(2)}$ (Mn^{II}–P $\rightleftharpoons$ Mn^{II}–P^{-})
Mn^{III}(OH) OEP	−0.42	−1.61
Mn^{III}(OH) ETIO	−0.45	
Mn^{III}(F) ETIO	−0.49	
Mn^{III}(Cl) ETIO	−0.45	−1.66
Mn^{III}(Br) ETIO	−0.41	
Mn^{III}(I) ETIO	−0.29	
Mn^{III}(Cl) TPP	−0.23	−1.38
Mn^{III}(Cl) $Pheo_a$	−0.23	−1.30
Mn^{III}(OH) Phth	−0.94	

[*Boucher* (*12*), *Lever*, (*122*), *Elvidge* (*53*)]

These data suggest that "hard-base" axial ligands, which form the strongest bonds to Mn^{III}, stabilize this oxidation state versus reduction. The same trend is found with the various porphyrin ligands where the porphyrins with the highest basicity stabilize the Mn^{III} by strong σ-electron donation from the nitrogens. Thus, even in the Mn^{III} porphyrins where the interaction between the metal and the porphyrin π electrons is extraordinarily strong as compared to other metalloporphyrins, the redox behavior of the central ion is determined by the sigma-donating power of the nitrogen atoms and not by the highly polarizable "soft" π electron conjugation system of the porphyrin ring, which would tend to favor low oxidation numbers [*Pearson* (*143*)]. All Mn^{III} complexes investigated so far are in the high-spin state, which also indicates the overwhelming strength of the in-plane porphyrin–metal interaction. Cyanide complexes could not be prepared.

The electrochemistry of the Mn^{II}/Mn^{III} couple has been studied in greater detail by the technique of cyclic voltammetry [*Kadish* (*108*)], which allows the study of the rate of electron transfer. The reduction of Mn^{III} is rather slow as compared to other metalloporphyrins, which could indicate a change in the ligand conformation, *e.g.* formation of a structure in which the manganese metal ion is considerably out of the porphyrin plane (see Mn^{II}). The reduction is also highly *p*H-dependent. The midpoint potential of Mn^{III} HEMATO in EtOH/H_2O is -350 V at *p*H 1, falls to a constant value of -550 mV between *p*H 4 and 10, and is further lowered to about -700 mV at *p*H 13 [*Davis* (*45*)].

Sublimation in vacuo of Mn^{III} porphyrins at high temperature also leads to quantitative reduction to the Mn^{II} porphyrins. This spontaneous reduction also takes place during measurement of the mass spectra [*Calvin* (*22*), *Edwards* (*52*)].

Another interesting observation concerns the spontaneous reduction of Mn^{III} to Mn^{II} porphyrins in pyridine solution in the absence of oxygen, which requires light [*Calvin* (*23*)]. Here reduction ceases and reoxidation occurs when the light is shut off, and no alteration of the porphyrin is found. This is one of the few reversible photoreactions of paramagnetic metalloporphyrins reported so far, and a more detailed study of the mechanism would probably yield information on the relationship between charge transfer to the axial ligands and possible internal redox reactions. Mn^{III} porphyrins, with their distinct charge transfer bands close to 800 nm and the wide variety of available axial anions, seem very appropriate for such studies.

Mn^{IV}. Chemical oxidation of Mn^{III} porphyrins at *p*H values above 9 yields a one-electron oxidation product that is thought to be a Mn^{IV} complex [*Calvin* (*23*)]. The magnetic moment is 2.0 BM which would roughly correspond to a low-spin d^3 case. Another possibility here would be the formation of a Mn^{III} cation radical, probable at lower *p*H. The strong nucleophilic environment in a highly basic solution may stabilize the $+4$ oxidation state of the central ion in the same way as it stabilizes the Mn^{III} oxidation state versus the Mn^{II} state.

Mn^{IV} porphyrins revert spontaneously to Mn^{III} porphyrins in solution and this process is accelerated by light [*Engelsma* (*54*)]; the stoichiometry of this reaction is not known. There has been speculation about the possible existence of an Mn^{V} porphyrin [*Calvin* (*23*)].

Similar light and dark "oxygen-breathing" reactions have been described with Mn—Phth, and the rich redox chemistry of this class of compounds has been reviewed [*Lever* (*122*)].

Rhenium

Re^{I}. A surprisingly acid-stable ($Re^{I}(CO)_3H$)-MESO complex with a "normal" electron spectrum [(585 (6,200), 485 (20,000), 392 (110,000)] has been obtained in crystalline form, but no magnetic data are available on this interesting, presumably non-planar d^6 system. A dirhenium complex ($Re^{I}(CO)_3)_2$-MESO (519, 400) has also been described [*Ostfeld* (*136*)].

Re^{V}. The diamagnetic Re^{V}O-OEP-fluoride produces an electronic spectrum [614 (3.76), 582 (3.92), 464 (4.64), 338 (4.81)], possibly indicative of an Mn^{III}-type system. The peak at 338 nm, or at least its high extinction, may not be real, however, because many of the authors' complexes have been measured in benzene or prepared in phenol so that aromatic compounds could largely account for this peak [*Buchler* (*21*)].

6. Group-VIII Porphyrins

Iron

Work on the redox chemistry of iron porphyrins and their protein adducts deals mainly with two problems, namely the electron transfer, which is mediated by the Fe^{II}/Fe^{III} couple in cytochromes, and the partial reduction of the oxygen molecule to the superoxide anion or to a peroxo bridge between two iron porphyrins. The work on isolated iron porphyrins will be summarized in this section, whereas the hemoproteins will be covered in Section VI.

Fe^{I}. The reduction of $(FeTPP)_2O$ in tetrahydrofuran by sodium amalgam yielded a light-red solution which produced a magnetic susceptibility μ_{eff} of 5.2 BM at room temperature ($S=\frac{3}{2}$) and an esr spectrum at 77 °K, which is indicative of an $S=\frac{1}{2}$ Fe^{I}-system ($g_{\parallel}=1.93$; $g_{\perp}=2.30$). The visible spectrum of this Fe^{I}(TPP) anion contains peaks at 540 nm (12,900) and 605 nm (sh, 4,200) [*Cohen* (*35a*)].

Fe^{II}. Crystalline Fe^{II} porphyrins are not easily available, because rapid autoxidation to Fe^{III} occurs even in the solid state. In contrast to the Mn^{II} porphyrins they have, however, been prepared under special conditions with the rigorous exclusion of oxygen. The procedure starts with the free-base porphyrin, which is chelated with Fe^{II} acetate in acetic acid. The resulting high-spin Fe^{II} porphyrin is crystallized directly from the reaction mixture with some water added under pure nitrogen [*Corwin* (*39*), (*40*), *Fischer* (*64*)]. *Corwin* reports "a peculiar golden lustre" of mesoheme in the solid state, which is not found in other metalloporphyrins [*Corwin* (*39*)]. The free Fe^{II} porphyrins are readily oxidized by oxygen, but addition of basic axial ligands, *e.g.* pyridine (hemochromes, see II.1) and absence of electron acceptors, *e.g.* water, leads to a considerable stabilization of the Fe^{II} state against oxygen and to diamagnetism of the central ion. Carbon monoxide also greatly stabilizes the Fe^{II} state [*Lemberg* (*121*)].

Alkaline aqueous solutions of hemochromes (see II.1) can be prepared from Fe^{III} porphyrins by a large variety of reducing agents (*e.g.* dithionite, ascorbic acid, sulfur dioxide) as long as a sufficient amount of base is present, *e.g.* the ratio of pyridine to heme should be at least 150,000:1 in the usual dithionite reduction [*Paul* (*140*)]. Under these conditions the typical hemochrome spectrum is obtained, with a strong, sharp α band at 558 nm and an equally sharp but less intense β band, even in the presence of oxygen. If the base is omitted, the ferroporphyrin is much less stable against oxidation and the visible spectrum becomes less intense and broad due to aggregation processes [*Falk* (*56*)]. A sample of carefully prepared heme esters in deoxygenated dioxane, however, also produces sharp spectra that are almost identical to the hemochrome spectra [*Corwin* (*39*)]. No magnetic measurements have been reported on this compound, although it is very probably of the high-spin state. One may rationalize this finding with the assumption that changes in the absorbtion spectra of metalloporphyrins with divalent central ions reflect their aggregation state rather than the spin state of the metal.

The stabilization of the d^6-Fe^{II} state through coordination to strong field ligands is accompanied by a change of the high-spin state in the aquo-iron^{II} porphyrins to low spin. The hemochromes are probably all diamagnetic and only in these cases is the redox chemistry known in some detail (see below). It is also thought that the high-spin five-coordinated complexes contain an out-of-plane iron, while the six-coordinated low-spin compounds are supposedly planar [*Falk* (*56*), *Gouterman* (*86*), *Hoard* (*94*)].

An interesting aspect of the redox activity of Fe^{II} porphyrins is their ability to reduce alkyl halides to hydrocarbon radicals, which then undergo various secondary reactions [*Wade* (*189*)]:

e.g. 1. $2\,RX + 2\,Fe^{II}\text{—DEUTERO} \rightarrow 2\,Fe^{II}\text{—DEUTERO—X} + R_2$
(R = ($C_6H_5CH_2$, $CH_2 = CHCH_2$, X = Cl, Br, J)

2. $(CH_2)_2 + 2\,Fe^{II}\text{—DEUTERO} \rightarrow 2\,Fe^{III}\text{—DEUTERO—X} + (CH_2)_2$

This indicates the highly nucleophilic character of the central electron-rich Fe^{II} ion, similar to the Co^{I} case (see IV.6).

Fe^{III}. The electronic spectra of high- and low-spin ferric complexes are usually of the "hematin" type and have been reviewed recently [*Willi-*

ams (*206*)]. Under strongly alkaline conditions, however, the equilibrium between $Fe^{III}(OH)(H_2O)P$ and $Fe^{III}(OH)_2$-P probably lies on the side of the symmetrically substituted dihydroxide in which the iron is forced into the porphyrin plane, and charge transfer from the porphyrin into the metal *d* orbitals is no longer important. Thus the spectrum is of the "normal" type (Fe^{III}—OH—OEP in dimethylsulfoxide with potassium tert. butylate: 548 nm) [*Horio* (*96*), *Salek* (*155*)].

Thus in all cases where an electronic spectrum of an iron porphyrin or of any other metalloporphyrin is not "normal", it is connected with an unsymmetrical axial coordination of the central ion and/or a non-planar coordination sphere of the metal ion in the porphyrin plane.

Recent esr measurements of low-spin Fe^{III} porphyrins at liquid helium temperature show sizable in-plane magnetic anistropy produced by the porphyrin skeleton ($g_1 = 2.9$, $g_2 = 2.3$, $g_3 = 1.5$) and the contact shifts observed in H—nmr spectra provided quantitative evidence for a $\pi \rightarrow d_{xz,yz}$ charge transfer [*LaMar* (*118a*)].

Ethyl-Fe^{III}ETIO has been prepared and is one of the rare examples of iron alkyl compounds [*Clarke* (*34*)]. The Fe^{III} oxidation state has also been extensively characterized by esr. X-ray, and Mössbauer, magnetic susceptibility and far-infrared data [*Blumberg* (*7*), *Wittenberg* (*210*), *Hoard* (*94*), *Lang* (*119*), *Haveman* (*90*), *Maricondi* (*127*), *Ogoshi* (*134a*)]. $Iron^{III}$ porphyrin dimers from the reaction

$$2\ Fe(OH)\ (H_2O)\text{—}P \xrightarrow{-H_2O} (Fe(H_2O)_2P)_2O$$

are also well-known and have been characterized [*Cohen* (*36*), *Wicholas* (*204*)].

Fe^{II}/Fe^{III} Redox Couple Iron Porphyrins

The midpoint potential of the hydrated inorganic ferrous-ferric couple at *p*H 7 is 530 mV and that for the heme-hematin reaction is in the range -450 to $+100$ mV. The general stabilization of the more highly oxidized state in the porphyrin field is an electrostatic charge effect of the porphyrin dianion: σ electrons from the nitrogens partly fill the $d_{x^2-y^2}$ orbitals of the central iron, thus reducing the positive charge from $+3$ to a value probably close to $+1$. Such an effect has been demonstrated strikingly with the phenanthroline iron complexes IX and X, where IX has a potential of 0.86 V (the divalent state is stabilized, because no anion can directly neutralize the positive charges on the metal ion), and the hexa-anion ligand in X (*p*H 10) lowers the Fe^{II}/Fe^{III} midpoint potential to $-0{,}34$ [*George* (*80a*)]. Thus, the rise in electrostatic charge around the iron atom leads to a drop in its oxidation potential of almost 30 kcal. It is the same general phenomenon that conditions the Fe^{3+}

oxidants to an energy range which is useful in biological systems by using porphyrins as ligands.

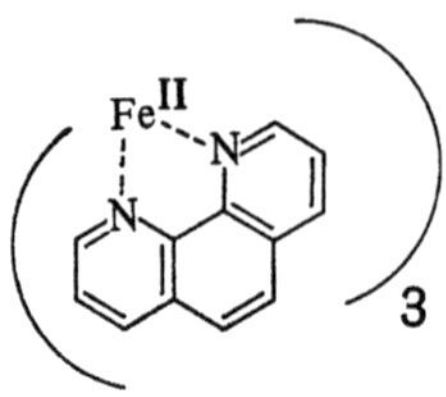

IX

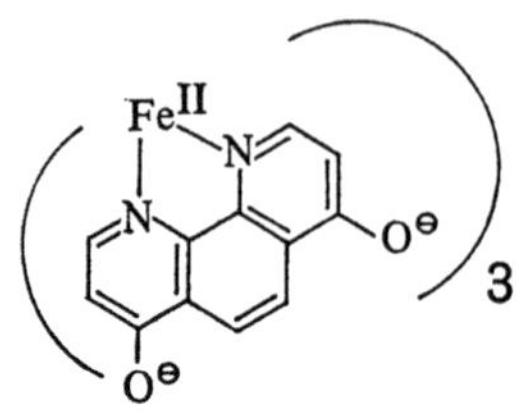

X

The redox behavior is further varied by factors introduced by the substituents of the porphyrin ring and the axial ligands, and work on these influences dominated the field of "redox behavior of metalloporphyrins" up to 1965 [*Clark* (*33*), *Falk* (*56*), *Lemberg* (*121*)].

From the above electrostatic argument it is clear that electron-donating substituents (increasing the negative ring charge and the porphyrin basicity) should lower the Fe^{II}/Fe^{III} potential, whereas electron-accepting substituents should raise it. This has been shown in many examples to be true, and Table 5 gives an important example where the effect of the two vinyl groups of protoporphyrin on the central iron in heme is substantiated [*Falk* (*56*)]. Tetraphenyl-porphyrins, chlorophylls and phthalocyanines all yield higher potentials than the corresponding prototypes of porphyrins, indicating that electrostatic interaction with the bound metal ion is less effective [*Davis* (*44*), *Falk* (*56*)].

Table 5. Oxidation potentials of Fe^{II} porphyrins in alkaline solution

Ferroporphyrin	$(Py)_2$complex; *p*H 9.6	$(CN^-)_2$; *p*H 9.6
MESO	−0.30	−0.47
HEMATO	−0.24	−0.44
PROTO	−0.22	−0.42

In the same way as porphyrin basicity influences the iron redox qualities, the basicity of the extra ligand plays an important role in this connection: the greater the donor power the more negative the potential. The relationship between potentials and the pk_a of various bases added to the central Fe^{II} ion (*e.g.* pyridine, nicotine, histidine, etc.) is essentially linear [*Barron* (*3*), *Falk* (*56, 57*)].

These simple, straightforward arguments indicate how nature tailors its redox catalysts: by variation of the porphyrin substitution pattern and, more important, by formation of mixed complexes with two different well-defined axial ligands from the protein core. This permits a wide variety of oxidation potentials. All experimental results indicate that the relative magnitude of the Fe^{II}/Fe^{III} redox potentials can be rationalized by simple electrostatic arguments and other considerations, *e.g.* of possible changes in the spin states or the covalency of the metal nitrogen bonds, are not needed. In order to understand biological electron transfer processes from the viewpoint of organic chemistry, it is, however not only necessary to trace the energy differences of the Fe^{II}/Fe^{III} couples, *e.g.* of different cytochromes, back to the iron porphyrins but also to form some idea of what produces the great differences in the rates of iron oxidation in such biologically important porphyrins in hemoglobin and cytochromes.

Voltammetric studies of electron-transfer rates of simple iron porphyrins and their dimers have shown that the transfer is very fast when both the Fe^{II} and Fe^{III} porphyrins are in the low-spin state (Fe-PROTO-$(CN)_2$: $k = 4.0$ cm/sec, EtOH/H_2O), slows down somewhat when both are in the high-spin state (Fe—H_2O—PROTO—OH: $k = 0.2$ cm/sec, EtOH/H_2O), and becomes very slow when a change from high-spin Fe^{III} to low-spin Fe^{II} is involved in the reduction (Fe—PROTO-Cl: $k = 8 \times 10^{-4}$, DMF). A dimerization of the porphyrin leads to a slightly higher reduction rate of the first partner [(Fe—H_2O—PROTO—OH$)_2$ $K = 0.8$, EtOH/H_2O)]. From these and other experimental results it seems well established that minimum solvation changes occur on reduction of the ferric porphyrins, even in a partial aqueous solution, when the spin state of the central iron remains unchanged, whereas a high-spin low-spin conversion dramatically slows down the electron transfer [*Kadish* (*108*)]. The biological implications of this general finding and the possible role of the porphyrin ligand in electron transfer will be discussed in Section VI.

The kinetics of autoxidation of various hemochromes were observed in twelve solvents of widely differing polarity, but the reaction turned out to depend upon the protic character and not the polarity of the solvent. The addition of oxygen and the oxidation of Fe^{II} to Fe^{III} in the solid state were also followed with various iron porphyrins. In pyridine-containing solutions the kinetics of autoxidations pointed to a mechanism that involved first the dissociation of a pyridine molecule from the initial dipyridine complex. The monomer pyridinate is attacked by an oxygen-bridged dimer. Rapid decomposition of the dimer yields Fe^{III} porphyrin hydroxides [*Cohen* (*36*)].

As a possible in-vitro model for one-electron transfer in photosynthesis, the photochemical reaction between hemin and Chl_a in pyridine solution has been studied, and it has been shown that the relatively slow reduction and oxidation of iron porphyrins can be accelerated by the presence of Chl_a by an order of magnitude and that light further increases the rate or reaction. Chl_a probably forms a rather stable complex with hemin, as is shown by fluorescence quenching experiments [*Brody* (*17*)].

The effects of concentration and sample purity on the redox potentials of the Fe^{II}/Fe^{III} couple in heme have also been studied: usually the midpoint potentials are much higher in concentrated solutions than in more dilute ones [*Adler* (*1*)]. Reduction of $Fe^{III}Cl^{36}$ TPP with Cr^{II} salts leads to a 96% percent transfer of the radioactivity towards the Cr^{III} product; which confirms the "inner-sphere" electron transfer across an Fe—Cl—Cr-bridge [*Cohen* (*37*)].

X-ray photoelectron spectroscopy (ESCA) of Fe^{II} $(piperidine)_2$ TPP and $Fe^{III}Cl$ TPP has produced a difference of 0.44 eV, in Fe $2p_{-3/2}$ binding energies, which has been tentatively interpreted that there is a difference of 0.44 electronic charges on the two types of iron. The negative charges on the nitrogens were found to be —0.13 in $Fe^{II}(pip)_2TPP$ and —0.04 in $Fe^{III}(Cl)$ TPP [*Zeller* (*218a*)]; see, however, II.1).

Fe^{IV}. Tetravalent heme iron has been claimed in many biochemical papers on heme enzymes that react with hydrogen peroxide, and recently an electrochemical preparation of Fe^{IV} TPP has been reported [*Brill* (*15*), *George* (*80*), *Iizuka* (*104*), *Felton* (*61*)]. The electronic spectrum of this compound is of the π-cation radical type [$(FeTPP)_2O$: Soret band 400 nm ($\varepsilon = 70{,}000$), one broad-structured band covering the whole visible range into the near infrared] and the total spin was shown to be $S = 2$. The π-cation radical, however, was not favored by the authors, since the oxidation product yielded a pmr spectrum where the phenyl proton signals were shifted by only 4 ppm and not the 20 ppm calculated for the $(FeTPP)_2O^+$ dimer, in which only one porphyrin ring is oxidized and fast electron exchange occurs with the unoxidized ring. The argument is based on the assumption that the observed hyperfine splitting constant of 0.3 G in the esr spectra of Zn TPP should be the same as in the Fe TPP dimer, with one of the TPP rings in the form of a cation radical. This, however, is by no means certain since the solvation sphere in a dimer will be different from that in a monomer, as will be the sterically hindered interactions between the phenyl and porphyrin rings. Other complications can be expected from different solvent–metal interactions with the unfilled d_{z^2} orbitals in Fe^{III}, as opposed to the unpolarizable d^{10} zinc ion. Moreover, the interaction between porphyrin radicals and connecting Fe—O—Fe bridges are not known. As long as this very special system of $(Fe\ TPP)_2O$ remains the only species in which the oxidation state "Fe^{IV}" can be studied, the alternative formulation of an Fe π-cation radical is just as likely for the first oxidation product of a Fe^{III} porphyrin. The normal voltammetric pattern of Fe^{III} porphyrins is also in agreement with both formulations [*Fuhrhop* (*74*), *Kadish* (*109*)].

Iron Porphyrin Oxygen Adducts

The biological implications and the redox chemistry involved in the reaction of oxygen with hemoproteins will be discussed in Section VI. Here only simple model systems which can reversible add oxygen will be briefly described.

In solution Fe^{II} porphyrins are rapidly oxidized to Fe^{III} by oxygen, but if at least a 1.5×10^5 excess of a nitrogenous base, preferably an imidazole derivative, is present, the reversible addition of oxygen is possible without any hematin formation [*Corwin* (*40*)]. Crystalline imidazole ferroprotoporphyrin has been prepared and 0.94 mol of oxygen could be reversibly bound to it [*Corwin* (*38*)]. The magnetic properties of this adduct are not known.

The addition of one molecule of oxygen to various iron porphyrin pyradinates has also been reported [*Alben* (*2*), *Caughey* (*25*)]. A mixture of phenylethylimidazole, polystyrene and Fe^{II}PROTO also forms stable oxygen adducts [*Wang* (*191*), *Wang* (*194*)]. No magnetic measurements or chemical reactivity studies have been performed on these oxyhemoglobin models.

Ruthenium

Ru^{II}. Ru^{II} TPP and MESO complexes have been described which produce "normal" electronic spectra [Ru—MESO—Im: 554 (85,500), 522 (69,500), 399 (114,000)] and are diamagnetic [*Chow* (*31*), *Cullen* (*42, 43*), *Tsutsui* (*178, 179*)]. The Ru^{II} porphyrins are rather air-stable in solution, and this stabilization of the lower oxidation state as compared to the iron porphyrins was explained as due to an assumed greater extent of π-bonding between the porphyrin and the larger d orbitals [*Chow* (*31*)], although the electronic spectra of Ru^{II} porphyrins do not offer any evidence for this.

Diamagnetic Ru^{II}OEP—CO is stable against air oxidation [$E^{ox}_{1/2} = 1.21$ V for the Ru^{II}/Ru^{III} couple and 0.64 V for the π-cation radical formation] and undergoes rapid dimerization when the carbon monoxide is split off by light. Preliminary conclusions from an X-ray study of $Ru^{II}(CO_2)$ TPP which implied bent R—C—O bonds (*154°*) [*Cullen* (*42*)] have been shown to be in error: The compound is in fact Ru^{II}(CO) (EtOH)—TPP and the Ru—C—O bond is linear (*176°*) [*Bonnett* (*8a*)]. Decarbonylated

Ru^{II} porphyrins are much more easily oxidized to Ru^{III} complexes ($E_{\frac{1}{2}}=0.6$ V) which have similar properties to the Fe^{III} porphyrins [*Brown* (*18*), *Sovocool* (*164*)].

Ru^{III}. The Ru^{III} TPP-dicyanide anion with a "normal" electronic spectrum [521 (7,200) 402 (120,000)] and one single electron ($\mu=2.1$ BM) is the only well-defined compound with this oxidation state [*Chow* (*31*)]. An early report has been shown to be in error [*Chow* (*31*), *Fleischer* (*69*)].

Osmium

Os^{II}. Only the diamagnetic Os^{II}OEP-carbonyl compound is known and yields a normal electronic spectrum [576 (1,200) 540 (12,000) 509 (8,000) 393 (155,000] [*Rohbock* (*152a*)].

Cobalt

The cobalt porphyrins deserve special interest because of their biological relevance to vitamin B_{12} chemistry, their capacity to bind oxygen reversibly and their catalytic activities in various redox reactions of the axial ligands. Both the latter aspects will be included in the following descriptions.

Co^{I}. The diamagnetic Co^{I} TPP has been obtained in solution by chemical reduction with various reducing agents (*e.g.* sodium borohydride, sodium amalgam) under nitrogen and by electro-chemical reductions [*Whitlock* (*201*), *Lexa* (*123*)].

The atypical electronic spectrum is reminiscent of a cation radical spectrum, and shows very broad visible absorption bands and a split Soret band of low intensity [608 (2,200) 511 (15,000) 427 (66,000) 362 (54,000)] [*Lexa* (*123*)]. Co^{I} ETIO reacts with acyl or aryl halides to form Co^{III} acyl or aryl adducts, which reminds one of the "supernucleophilic" catalytic activities of vitamin B_{12}:

$$Co^{I}\,ETIO(H) + CH_3J \rightleftharpoons Co^{III}\,(CH_3)\,ETIO(J)$$

[*Schrauzer* (*155a*)].

Co^{II}. These d^7 metalloporphyrins produce the expected magnetic moment of $1\cdot8$ BM in pyridine solution and show normal electronic spectra [*Havemann* (*90*)].

Esr spectra of Co^{II}–TPP complexes with different axial ligands have been studied in detail in toluene glasses, and the addition of one oxygen molecule in the presence of air seems probable:

$$\textit{e.g.}\ Co\,(amine)\text{—}(p\text{-}OCH_3)TPP + O_2 \rightleftharpoons Co\,(amine, O_2) - (p\text{-}OCH_3)TPP$$

[*Walker* (*190*)].

From the spectra (Fig. 4) it was concluded that an electron is transferred from the d_{z^2} orbital of cobalt to a π^* orbital of the oxygen molecule. The original signal of the Co^{II} was restored to its former intensity by

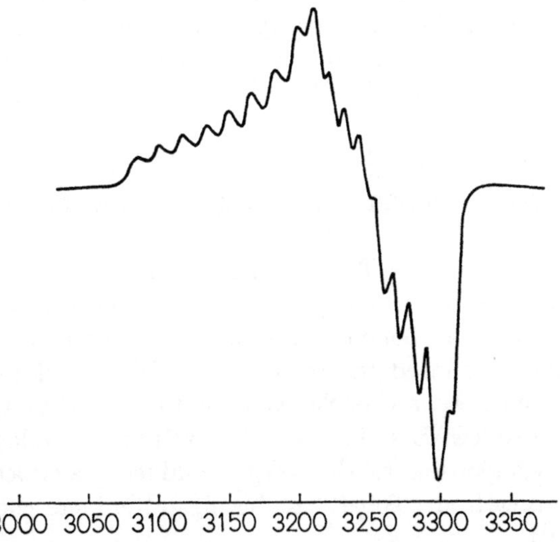

Fig. 4. Esr spectrum of the O_2 adduct of Co TPP (*Walker, 190a*).

pumping off the oxygen, indicating that the formation of superoxide complexes is fully reversible:

$$Co^{II} + O_2 \rightleftharpoons Co^{III}\,O_2^{\ominus} \longleftrightarrow Co^{II} - O_2$$

Later, dimethyl formamide solutions of Co^{II} MESO were shown to add oxygen reversibly. The solvent effect [*Stynes* (*170*)], the thermodynamics of oxygen binding to Co^{II}porphyrins with various bases [*Walker* (*190a*), *Stynes* (*170a*)], as well as the kinetics [*Stynes* (*170b*)] have been studied and the formation of peroxo-bridged complexes appears probable (*170b*). No chemical proof of superoxide or peroxide reactivities of the bound oxygen has been reported so far in these or related systems. If Co^{II} porphyrins are placed in solvents with high dielectricity constants, they are rapidly oxidized to Co^{III} [*Tsutsui* (*180*), (*182*), *Whitten* (*202*)].

Addition of olefins to a solution of Co^{II} Meso results in an immediate shift of the Co^{II} 393-nm Soret band to the Co^{III} 411-nm absorption.

Various hydrocarbons have been used in this reaction, but no product analysis has been performed on any of them [*Tsutsui* (*180*), (*182*)]. These experimental results indicate efficient activation of molecular oxygen, and the intermediacy of superoxide and/or peroxo complexes of the cobalt porphyrins is probable.

Co^{II}TPP reacts with tetracyanoethylene (TCNE) to a 1:1 complex, where nitrogens of the nitrile groups interact with the cobalt (Eqn. 1). If pyridine is added to this complex, together with an excess of TCNE the molecule of TCNE is reduced by the Co^{II} to the anion radical (eqn. 2).

$$Co^{II}TPP + TCNE \rightleftharpoons Co^{II}TPP \cdot TCNE \text{ (N-bonded)} \quad (1)$$

$$Co^{II}(Py)(TCNE)TPP + TCNE \rightleftharpoons [Co^{III}(TCNE)(Py)]^{+}TCNE^{-} \quad (2)$$

The esr signal of the TCNE radical does not show any hyperfine structure produced by coupling to the cobalt nucleus, which indicates that it is not directly bound to the metal. The presence of the strong donor pyridine enhanced the fraction of Co^{II}TPP oxidized by TCNE (eqn. 2) and complete, reversible oxidation to [Co^{III} (Py) TPP]$^{+}$ TCNE^{-} (or its N-bonded TCAE adduct) could be achieved in toluene [*Wayland* (*197a*)]. In comparison to the oxygen adducts described above, the electron transfer from "cobalt" to the ligand is more complete in the TCNE adduct.

Co^{II}TPP also reversibly forms adducts with sulfur dioxide which, in contrast to oxygen, functions as an electron donor [*Wayland* (*197*)]. Evidence for this statement comes from the reduction of the $g_{\perp}$ and the ^{59}Co coupling constants in the esr spectrum of the Co^{II} (SO_2) TPP adducts as compared to the values for CoTPP. The observed behavior was taken as characteristic of donor adduct formation and assumed to result primarily from an increase in the $d_{xz,yz} \rightarrow d_{z^2}$ energy separation.

Co^{III}. The diamagnetic Co^{III} porphyrins are prepared from Co^{I} of Co^{II} complexes, as described above. The electronic spectrum is always "normal", but the ratios of the visible α and β bands, as well as the exact position of the Soret band are solvent-dependent to an exceptional degree. The oxidation $Co^{II} \rightarrow Co^{III}$ has been shown to be rapidly reversible on changes of temperature and solvent, but the exact nature of the reduction of Co^{III} porphyrins, which sometimes even occurs in air, is not known [*Whitten* (*202*)]. Co^{III}(Br)(Py)ETIO reacts with a variety of alkyl and aryl magnesium bromides to give the corresponding alkyl and aryl cobalt^{III} derivatives:

$$Co^{III}(Br)ETIO + CH_3MgJ \rightarrow Co^{III}(CH_3)\,ETIO + MgBrJ$$

[*Clarke* (*34*)].

Visible spectra and nmr data point to the correctness of the assignment of the Co^{III} oxidation state in the product.

Rhodium

Rh^I. The diamagnetic H^+ $(Rh_2(CO)_4$ OEP $C1)^-$ contains a Rh—Cl—Rh bridge and has been prepared in the form of green crystals [*Yoshida* (*218*)]. Its electronic spectrum resembles that of the Co^I porphyrins [367 (66,000) 444 (39,000) 516 (13,800) 590 (5,100)]. In contrast to the Co^I complexes, however, the Rh^I complexes are rather air-stable and have been chromatographed on silica gel, where they convert to the Rh^{II} complex only after several hours of air exposure.

A new type of stable metalloporphyrins, recently introduced by *Yoshida*'s group, produces a very interesting kind of intramolecular redox chemistry. $[Rh^I(CO)_2Cl]_2$ reacts with N-methyl-octaethylporphyrin to a "sitting-atop" complex (XI) [617 (3,500) 595 (5,500) 570 (6,000) 555 (5,000) 430 (30,000) 398 (36,000)] which rearranges on gentle heating to Ru^{III} $(CH_3)(H_2O)_2$ OEP (XII) [544 (32,500) 512 (20,000) 396 (129,000) 384 (95,500)] [*Ogoshi* (*134b*)].

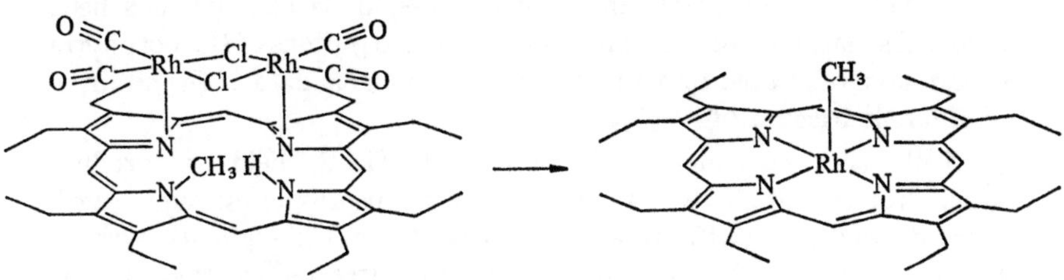

XI XII

The exact stoichiometry and the mechanism of this reaction are not yet known. Compound XI will probably react with other electrophiles and might help to introduce a variety of organic residues into the porphyrin center through rearrangement to Ru^{III} porphyrins.

Rh^{II}. The TPP complex has a normal electronic spectrum (601, 568, 531; 418; III < II < I), a magnetic moment of approx. 1.2 BM, and an anisotropic esr spectrum ($g = 2.089$, 2.029, 1.990, $g_{av} = 2.036$). It is reduced by molecular hydrogen to a diamagnetic Rh^I porphyrin hydride. which is readily reconverted to the Rh^{II} complex by molecular oxygen [*James* (*107*)].

Rh^{III}. $Rh^{III}(Cl)(H_2O)_2$ OEP is also diamagnetic and produces a normal spectrum [403 (132,000) 520 (13,200) 554 (25,000)]. The reaction of this complex with lithium alkyls led to alkylated rhodium compounds, whereas the Grignard reagents RMgX simply replaced the chlorine anion by X [*Yoshida* (*218*)]. A preparation of Rh^{III} Meso showed weak residual paramagnetism (0.4 to 0.5 BM) and a free-radical esr spectrum due to less than 0.1% of the sample [*Sadavisan* (*154*)]. It seems possible that the "residual" susceptibility indicates some Rh^{II} porphyrins.

Iridium

Ir^{III}. Only the diamagnetic Ir^{III} porphyrins are well characterized [Ir(CO-MESO (Cl) 548, 515, 398], although the spectroscopic occurrence of Ir^{I} porphyrins has also been shown [*Sadavisan* (*154*)].

Nickel

Ni^{II}. This d^8 system is one of the most stable oxidation states in metalloporphyrins, and the nickel complexes produce normal, solvent-dependent electronic spectra. In the usual organic solvents Ni^{II} porphyrins are diamagnetic, but on addition of nitrogenous bases a magnetic moment of 3.45 BM was found [*McLees* (*133*)], which points to an octahedral ligand field. No esr spectrum of this high-spin d^8 complex has been reported so far, but contact shifts of the methine protons of the porphyrin ligand have been used to obtain the thermodynamic data of the addition of different bases (*133*).

Ni^{III}. Electrochemical oxidation of Ni^{II}TPP yields a product which produces at 77 °K a slowly decaying anisotropic esr signal with $g_{\perp}=2.30$ and $g_{\parallel}=2.12$, which is typical for a high-spin d^7 system. This esr spectrum changes within hours to a free radical spectrum at liquid-nitrogen temperatures [*Wolberg* (*211*)]. No explanation of this behavior has been given. No other supporting evidence is available for the supposed Ni^{III} porphyrin from chemical or electrochemical work with TPP or other nickel porphyrins. This interesting oxidation state is certainly a good candidate for more detailed work.

Palladium, Platinum

Pd^{II}, Pt^{II}. Both metal ions produce diamagnetic porphyrin complexes with normal electronic spectra. Very often they contain free radical impurities, probably acquired during complex formation with the strongly oxidizing Pd^{II} and Pt^{II} salts. With low-potential porphyrin ligands (*e.g.* certain chlorophyll and porphyrin derivatives with electron-donating groups on the methine bridges) the ligand may exist to a high degree in its free radical form [*Fuhrhop* (*70*)].

7. Group-Ib Porphyrins

Copper

Cu^{II}. The magnetic moment of Cu^{II} porphyrins points to a d^9 system (1.64 BM) and the normal electronic spectrum to an undisturbed porphyrin π system. Extensive esr studies are available which, in solution, show fine splitting through the copper nucleus and hyperfine splitting through the porphyrin nitrogens ($g_{\parallel}=2.16$ $g_{\perp}=2.00$) [(*Blumberg* (*6*), *Roberts* (*151*)]. No other oxidation states of copper have been reported.

Silver

Ag^{I}. A silver^{I} salt of TPP containing two silver atoms has been described; it produces an electronic spectrum like that of the porphyrin dianions [*Rothemund* (*153*)]. Ag_2^{I} phthalocyanine is photolabile [*Lever* (*122*)].

Ag^{II}. This is the usual oxidation state of silver in the porphyrin cavity and is obtained from silver^{I} salts and the free porphyrin bases. Its electronic spectrum is normal [Ag OEP 560 (18,000) 526 (12,500) 409 (219,000)] and its esr signal can be explained by the superposition of sets of nine lines produced by the hyperfine splitting through the nitrogens ($g_{\parallel}==2.10$ $g_{\perp}=2.03$) [*Kneubühl* (*114*)].

Ag^{III}. The chemical oxidation of Ag^{II} OEP leads to Ag^{III} OEP salts, which are stable, diamagnetic and produce a normal electronic spectrum [522 (25,200) 516 (9,800) 404 (131,000)] [*Kadish* (*109*)].

Gold

Au^{III}. Au^{III} TPP salts have been prepared by various groups; they are presumably diamagnetic and produce a "normal" electronic spectrum [Au TPP 521 (20,000) 412 (330,000)] [*Fleischer* (*66*)]. No other oxidation states are known with porphyrins, but Au^{II} Pht has been identified by its esr spectrum [*MacCragh* (*125*)].

8. Group-Ia, IIa-IIb, and -IIIb Porphyrins

Ia porphyrins are salts with two M^+ counterions; IIa and IIb porphyrins contain metals in their +2 oxidation states; IIIb porphyrins in the +3 state. All complexes are diamagnetic and produce normal electronic spectra. Most of them fluoresce. For thallium a labile green TPP complex has been described (696, 646, 483, 458), which could be the Tl_2^{I} salt [*Rothemund* (*153*)]. A report on Tl^{II} porphyrins (*153*) is probably in error, because in air only Tl^{III} porphyrins have been found in later investigations [*Smith* (*162*)]. Tl^{III} in trifluoroacetic acid also oxidizes porphyrins and chlorins in good yield to oxo-phlorins and bile pigments [*Cavaleiro* (*26*)].

9. Group-IVb Porphyrins

Silicon, Germanium

Si^{IV}, Ge^{IV}. Only the four-valent state of these atoms is known with porphyrins. The electronic spectra are "normal" [Si-ETIO-$(OH)_2$ 572 (13,800) 535 (11,700) 405 (314,000)] [*Boylan* (*13*), *Kane* (*110*)]. [Ge-OEP-$(OH)_2$: 570 (12,000), 534 (11,200) 404 (280,000)] [*Fuhrhop* (*70*)].

Tin

Sn^{II}. If Sn^{II} salts are reacted with porphyrins one usually obtains Sn^{IV} porphyrins, but in rigorously dried and degassed pyridine another complex is obtained with a Soret band close to 460 nm [*Whitten* (*203*)]. This could not be demetalated, neither could the metal be exchanged with copper salts. The latter treatment results, however, in formation of Sn^{IV} porphyrins. Reduction of Sn^{IV} porphyrins with sodium boro-hydride leads to tin porphyrins which can be demetalated by hydro-chloric acid, but no 460 nm band is found in the reaction mixture [*Fuhrhop* (*75*)]. Thus, the properties of Sn^{II} porphyrins remain unconfirmed.

Sn^{IV}. These metalloporphyrins are extremely acid-stable and produce a "normal" electronic spectrum [Sn^{IV} OEP $(OAc)_2$: 576 (19,100) 538 (19,100) 411 (403,000)] [*Parnemann* (*139*)]. The electrochemical redox pattern probably does not involve a $Sn^{IV} \rightleftharpoons Sn^{II}$ reduction. It follows that the porphyrin ligand is reduced, not the Sn^{IV} central ion [*Fuhrhop* (*74*)].

Lead

Pb^{II}. These metalloporphyrins are extremely acid-labile and produce the typical "lead–porphyrin" electronic spectrum with a Soret band close to 470 nm (see text relating to Fig. 1).

Pb^{IV}. A complex has been obtained by electrochemical oxidation of the Pb^{II} complexes. This oxidation is the only case where the porphyrin ligand is first oxidized to a cation radical, which then on further oxidation converts to the porphyrin with a four-valent central ion

$$Pb^{II}\,OEP \underset{}{\overset{-e}{\rightleftharpoons}} Pb^{II}\,OEP^{+} \overset{-e}{\rightleftharpoons} Pb^{IV}\,OEP$$

[*Ferguson* (*62*)].

Pb^{IV} porphyrins are also formed on chemical oxidation, have a normal electronic spectrum (*577*, *538*, *409*) and are extremely unstable [*Fuhrhop* (*74*)].

10. Group-Vb Porphyrins

The As, Sb, and Bi complexes of MESO have been prepared but except for Sb^{III} the oxidation states are not known with certainty [*Treibs* (*175*)]. Besides the complexes with +3 central ions, which produce "normal" spectra [Sb^{III} MESO **574**, **536**], other spectral types are found, probably belonging to the +5 oxidation state. [*Treibs* (*175*)].

11. Summary

All known oxidation states of metalloporphyrins are shown in Table 6. The main effect of the porphyrin ligands on central metal ions is to stabilize relatively high oxidation states [Fe^{III}, Mn^{III}, V^{IV}, Sn^{IV}] through strong σ donation by the porphyrin ligand. Redox data of Fe^{III} and Mn^{III} porphyrins with different porphyrin ligands show that the high-valent states are most stabilized when the porphyrins contain strong electron-donating substituents and when the hard-base axial ligands are added. The large conjugated system of the porphyrins does not have any clearcut direct effect on the redox behavior of the central ions.

Another strong and specific influence of the porphyrin field comes from its usually rigid planarity. Square planar ligand fields tend to stabilize d^8 transition-metal ions which, after the donation of σ electrons from the porphyrin ligand, approach the particularly stable d^{10} configuration, *e.g.* Co^I, Rh^I, Ni^{II}, Ag^{III}. Extended Hückel calculations on copper porphyrins have predicted for the same reason a high-energy electron in this d^9 system, so that oxidation should lead to diamagnetic d^8 Cu^{III}, but this has been shown experimentally to be wrong. Only the large Ag^{II} ion produces this effect [*Zerner* (*219*), *Kadish* (*109*)]. The same stabilizing effect, to a lesser extent, can be found with the "half-filled" d^4 systems [Cr^{II}, Mn^{III}, Fe^{IV} (?)]. From these typical ligand-field arguments it becomes clear that the porphyrin ligands act towards their central ions as a "hard" base although they contain a highly polarizable "soft" π-electron core. The overwhelming strength of the in-plane porphyrin–metal interaction is also evidenced by the fact that very few metals and only very strongly bound axial anions give rise to low-spin complexes, *e.g.* no anion with Mn^{III} and CN^- with Fe^{III} [*Pearson* (*143*), *Boucher* (*11*)]. (see also section II_3).

Table 6. The known oxidation states of Metalloporphyrins

The Roman numbers give the oxidation numbers found for the central ions, the π signs indicate the oxidation states (except zero) of the porphyrin π systems investigated so far. Brackets indicate doubtful cases. Phthalocyanines are not included.

Li I	Be II											B	C	N	Ne		
Na I, ($\pi^{\cdot -}$)	Mg II $\pi^{\cdot +}$, π^{2+}, $\pi^{\cdot -}$											Al III	Si IV	P	A		
K I	Ca II	Sc III	Ti IV	V IV	Cr II, III, (IV)	Mn II, III, (IV)	Fe I, II, III, (IV) ($\pi^{\cdot +}$)	Co I, II, III, $\pi^{\cdot +}$	Ni II, III, $\pi^{\cdot +}$	Cu II, $\pi^{\cdot +}$	Zn II, $\pi^{\cdot +}$, π^{2+}, $\pi^{\cdot -}$, π^{2-}	Ga III	Ge IV	As (III)	Kr		
Rb I	Sr II	Y	Zr IV	Nb V	Mo IV, V	Tc	Ru II, III, $\pi^{\cdot +}$	Rh I, II, III	Pd II, $\pi^{\cdot +}$	Ag II, III	Cd II	In III	Sn IV, (II)	Sb III	Xe		
Cs	Ba II	La	Hf IV	Ta V	W V	Re I, V	Os II	Ir III, (I)	Pt II	Au III	Hg II	Tl (I), III	Pb II, IV	Bi (III), (V)	Rn		
Fr	Ra	Ac															

The σ-donating power of related ligands decreases in the order ETIO > TPP > Pheo a > Pht.

Up to an ionic radius of at least 0.95 Å (*e.g.* Tl^{III}) the porphyrin ligands form stable complexes with metal ions by the indicated donation of σ electrons from the central nitrogens. With ions larger than one angstrom the bond will become more ionic and the porphyrin ligand may become bent (*e.g.* Pb^{2+}).

These simple qualitative arguments suffice to rationalize all the exceptional oxidation states given in Table 6 (*e.g.* Cr^{II}, Mn^{III}, Co^{I}, Rh^{I}, Ag^{III}, *no* Sn^{II}] and the numerical values of the potentials in Table 7. A more detailed analysis will not be possible until we have new and more precise models because the existing ligand-field and extended Hückel calculations do not reproduce correctly the relative energies of metal-d and porphyrin-π orbitals.

The simplicity of the redox patterns produced by metal ions in the porphyrin cavity also renders superfluous the various models for porphyrins which, being usually not very rigid and unsymmetrical macrocycles, tend to complicate rather than clarify the situation [*Tang* (*172*), *Tokel* (*174*)].

V. The Oxidation States of the Porphyrin Ligands

1. General Description

The potentials of reversible one-electron oxidations (reductions) of alternant aromatic hydrocarbons have been found to give good straight lines when plotted against the calculated energies of the highest occupied molecular orbitals (HOMOs; lowest unoccupied molecular orbitals = LUMOs) [*Heilbronner* (*91*)]. The results indicate two important trends:

1. The more extended the conjugated system, the easier it is to oxidize and reduce; HOMOs and LUMOs are always equally far apart from the nonbonding level with different signs in the relative energies (conjugative effect).
2. Electron-donating substituents (*e.g.* $-O^-$) lower the oxidation potential, but also lower the reduction potential to more negative values (inductive effect).

Thus, the results of redox experiments on conjugated hydrocarbons can give good experimental evidence for the energies of the outer orbitals of related conjugated systems in respect to a reference compound, and so the nature of substituent effects can be evaluated. The difference in energy between HOMOs and LUMOs can be roughly determined from the energy of the excited state, *e.g.* through the electronic spectra or, independently, from the difference $E_{1/2}^{ox}$—$E_{1/2}^{red}$ in ground-state redox experiments. [*Heilbronner* (*91*), *Hoijtink* (*95*)]. Finally, the square of the coefficients of the HOMOs (LUMOs), which are identical to the calculated unpaired electron densities in the same orbitals at the various centers of the molecules, can be obtained from the hyperfine splitting spacings, if any, in the esr spectra of the cation (anion) radicals. Surprisingly, the agreements between simple Hückel model calculations, more sophisticated treatments, and experimental results are quite good.

In theoretical interpretations of their electronic [*Gouterman* (*85*), *Simpson* (*160*), *Weiss* (*199*)] and pmr spectra [*Jackman* (*105*)] porphyrins have often been treated as cyclic polyene hydrocarbons, and it has recently been demonstrated that the (16)-annulene π dianion (which has 18 π electrons, is planar and aromatic) does indeed perfectly reproduce the spectra of porphyrin species with D_{4h} symmetry [*Oth* (*137*)]. The analogy between aromatic hydrocarbons and the porphyrins stands on solid experimental and theoretical ground, and redox potentials together with any coupling constants from esr experiments should yield equally straightforward correlations. Zeeman effect measurements also support this conclusion [*Malley* (*126*)].

The main interest with porphyrins, however, lies not in the effects of variations in the conjugated π system but in the influence of various metal ions in the porphyrin cavity on the redox activity of the porphyrin ligands. With transition-metal ions both conjugative and inductive effects can be expected, and regularities in the oxidation and reduction potentials, as well as the differences between them, should yield direct information on the nature of the metal–porphyrin interactions.

The experimental fundamentals for differentiating between metal and porphyrin ring reactions have been given in Section III, thus only the products of the "organic" redox reactions will be discussed in the following paragraphs.

A systematic voltammetric investigation of the redox potentials of 23 different metal complexes of OEP including all members of the first transition series (Table 7) has yielded some electrochemical regularities [*Fuhrhop* (*74*), *Kadish* (*109*)]:

1. The differences in potential between the first reduction and oxidation of the porphyrin ring are, with two exceptions (Mn, Mo), constant (2.25 + 0.15 V).

2. The differences in potential between the removal of the first and second electron are constant (0.3 ± 0.1 V).

3. The intervals between the potentials necessary to add one and two electrons to the porphyrin ligand are also constant (0.4 ± 0.1 V). These empirical rules together with chemical and spectroscopic data have

been used to correlate the various redox steps shown in Table 7 with the reactions indicated. The compliance of Fe^{III} OEP with all of the above rules should be mentioned. It is not possible to extract from these data any support for a hypothetical Fe^{IV} state (see IV.6). The exceptions of Mn^{III} and Mo^{V} porphyrins have been tentatively related to the exceptional $d_{xy,yz}$—π interaction in these metalloporphyrins (see IV.5). With stable divalent central ions, linear correlations between redox potentials and metal electronegativities have been demonstrated [*Fuhrhop* (*70a*)].

The most surprising and interesting of these electrochemical rules is certainly the first one: regardless of any intermediate change in the oxidation states of the central ions, the difference between the highest occupied and lowest unoccupied orbitals is a constant 2.2 eV in the ground state, which compares favorably with Gouterman's calculated value of 2.17 eV [*Zerner* (*219*)]. Both the ring oxidation and reduction potentials can

Table 7. Summary of Redox Potentials for octaethyl and Metallo-octaethyl porphyrins $E_{\frac{1}{2}}$'*s* in volts vs. S.C.E.

	Ligand	Oxidation	Metal	Ligand	Reduction
	$E_{\frac{1}{2}}(2)$	$E_{\frac{1}{2}}(1)$	$E_{\frac{1}{2}}$	$E_{\frac{1}{2}}(1)$	$E_{\frac{1}{2}}(2)$
H_2OEP	1.30	0.81		−1.46	−1.86
H_4OEP^{+2}		1.65			
Mono-N-MeOEP		1.37		−0.73	−1.17
Mono-N-MeOEP⁻		0.86		−1.37	
AgOEP		1.10	0.44 III ⇌ II	−1.29	
Al(OH)OEP	1.28	0.95		−1.31	
CaOEP	0.86	0.50		−1.68	
CdOEP	1.04	0.55		−1.52	
CoOEP		1.00	Irreversible III ⇌ II −1.05 II ⇌ I		
Cr(OH)OEP	1.22	0.99	0.79 IV ⇌ III −1.14 III ⇌ II	−1.35	
CuOEP	1.19	0.79		−1.46	
Fe(OH)OEP	1.24	1.00	−0.24 III ⇌ II	−1.33	

Table 7 (continued)

	Ligand Oxidation		Metal	Ligand Reduction	
	$E_{\frac{1}{2}}(2)$	$E_{\frac{1}{2}}(1)$	$E_{\frac{1}{2}}$	$E_{\frac{1}{2}}(1)$	$E_{\frac{1}{2}}(2)$
Ga(OH)OEP	1.32	1.01		−1.34	−1.80
$Ge(OH)_2OEP$	1.36	1.09		−1.31	
In(OH)OEP	1.36	1.08		−1.19	−1.59
MgOEP	0.77	0.54		−1.68	
Mn(OH)OEP	1.4	1.12	−0.42 III ⇌ II	−1.61	
MoO(OH)OEP		1.43	−0.21 V ⇌ IV	−1.30	−1.72
NiOEP		0.73		−1.5	
PbOEP		0.65	0.9 II ⇌ IV	−1.30	
PdOEP		0.82		−1.53	
Sb(OH)OEP		1.4		−1.07	
Sc(OH)OEP	1.03	0.70		−1.54	
$Si(OH)_2OEP$	1.19	0.92		−1.35	
$Sn(OH)_2OEP$		1.4		−0.90	−1.30
TiOOEP	1.32	1.03		−1.21	−1.69
Tl(OH)OEP	1.31	1.00		−1.24	
VOOEP	1.25	0.96		−1.25	−1.72
ZnOEP	1.02	0.63		−1.61	

vary by as much as one volt with different metal ions (*e.g.* Mg, Sn^{IV}) and the fact that a rise in oxidation potential is accompanied by a lowering of the reduction potential clearly points to strong inductive effects of the metals on the π system of the porphyrin ligand.

Although the major porphyrin–metal interaction now seems established as a simple σ-electron donation from the porphyrin nitrogens to the metal, the chemical consequences to be drawn for the π-electron system from the porphyrin ring are by no means trivial: whereas Mg OEP can be oxidized quantitatively by a small excess of iodine to the cation radical [*Fuhrhop* (*76, 78*)]:

$$2\,\mathrm{Mg\,OEP} + I_2 \underset{K1}{\rightleftharpoons} 2\,\mathrm{Mg\,OEP}^{+} + 2\,I^{-},$$

Sn^{IV} OEP is fully stable against bromine [*Fuhrhop* (*75*)]:

$$Sn^{IV}(OAc)_2OEP + Br_2 \longrightarrow \text{no reaction.}$$

whereas Mg OEP and Ca OEP are the only OEP metal chelates which can be photooxygenated with molecular oxygen (XIII→XV):

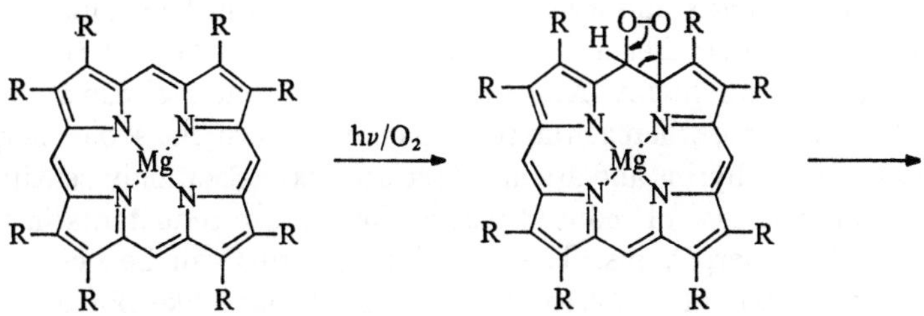

XIII XIV

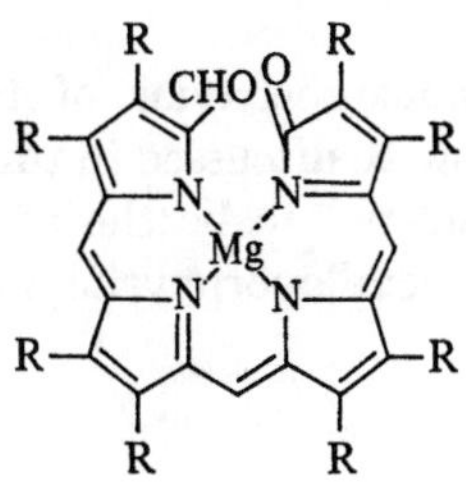

XV

R = –C_2H_5

[*Fuhrhop* (*77*), *Wasser* (*195*)], Sn^{IV} OEP (XIV) and Ge^{IV}OEP are uniquely reactive in a mild photoreduction procedure to phlorins (XVII) and chlorin (XVIII) [*Fuhrhop* (*75*)]:

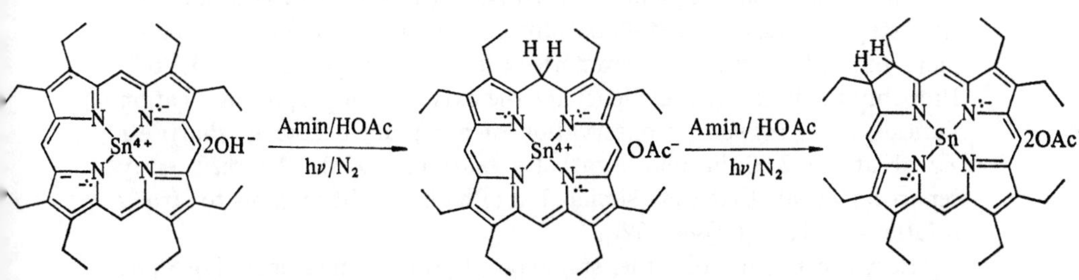

XVI XVII XVIII

Many other reactions of the porphyrin ligands have been found to be highly dependent on the central ions, and the correlations are usually straightforward. Some aspects of irreversible porphyrin ligand reactions have been covered in recent reviews on the chemistry of porphyrins [*Hambright* (*88*), *Inhoffen* (*99*), *Inhoffen* (*100*), *Smith* (*161*)].

Oxidation of the porphyrin ring is sometimes followed by reversible addition reactions of nucleophiles [*Dolphin* (*50*)]; reduction of the porphyrin is often accompanied by reversible addition reactions of electrophiles [*Peychal-Heiling* (*147*, *148*)]. Both reactions occur invariably on the methine bridges and lead to phlorins, which can be compared to the σ complexes of smaller aromatic rings, except for the fact that the latter complexes are highly reactive and under normal circumstances have a short lifetime, whereas the phlorins have almost the same energy content as the porphyrins and are in a first approximation only sensitive to oxidants in the case of reduced porphyrins, and to reductants in the case of oxidized porphyrins. The metalloporphyrins can be described as aromatic compounds, whereas the phlorins behave like polyenes in that the π system of the former is much more stable and undergoes reversible reactions, and the latter is more likely to react irreversibly to more saturated compounds.

The various oxidation states and the chemical behavior of the porphyrin ligands in their metal complexes will now be discussed in turn. The π-cation radicals and the phlorins are considered to be the most important oxidation and reduction products of metalloporphyrins and will therefore be described in more detail.

2. π-Cation Radicals and π Dications

Rabinowitch (*150*) reported early studies on the oxidation of chlorophyll with Fe^{III} salts and found dramatic, reversible spectrophotometric changes, later reproduced with magnesium tetrabenzoporphyrin [*Goedheer* (*83*, *84*)]. *Stanienda* demonstrated polarographic one-electron oxidation steps in various metalloporphyrins where the metal was redox-inert, but did not further characterize the oxidation products (*165*, *167*). Probably the first direct evidence for the existence of porphyrin π cation radicals was *Commoner*'s observation of a rapidly decaying, sharp esr signal at $g = 2.00$ in photosynthetic material (*37a*). Another early report on signals that were similar but of high stability relates to strong solutions of TPP [*Gibson* (*82*)].

Easy accessibility of rather stable porphyrin cation radicals, however, as well as their detailed analysis, was only achieved when esr spectra, absorption spectra and redox-potentials had been quantitatively related to the oxidative formation of these radicals and crystalline material had been obtained.

Magnesium Porphyrin Cation Radicals

The oxidation of magnesium porphyrins in organic solvents to their cation radicals were studied in great detail because of their interest for some reactions of the chlorophylls in photosynthesis. Chemical or electrochemical oxidation of various magnesium porphyrins yields stable, crystallizable π-cation radicals and is fully reversible [*Fuhrhop* (*76, 78*)].

A comparison of the midpoint potentials of some typical complexes (Table 8) shows that the potentials of metallochlorins lie about 300 mV

Table 8. Oxidation potentials of some magnesium porphyrins [mV]

Ligand	$E^{Ox}_{\frac{1}{2}}(1)$	$E^{Ox}_{\frac{1}{2}}(2)$	Literature
OEP	427		(*78*) *Fuhrhop*
PROTO	520	940	(*165*) *Stanienda*
TPP	540	860	(*165*) *Stanienda*
OEC	107		(*70a*) *Fuhrhop*
Chl_a	550	776	(*78, 166*) *Fuhrhop, Stanienda*
BChl	270		(*78*) *Fuhrhop*

lower than those of analogous porphyrins (compare OEP and OEC), and bacteriochlorins are again 300 mV lower (compare Chl_a and BChl). Metal complexes of bacteriochlorins derived from ETIO-type porphyrins should therefore have very low oxidation potentials (*e.g.* Mg-octaethylbacteriochlorin −200 mV). Such compounds, however, could not be prepared so far. This failure can be rationalized with the arguments given in Section II.2 and II.3: in the β-hydrogenated porphyrin ligands one or two of the electron-attracting pyrrole units, which can be formulated as pyrrole anions (see II.3; $pK \sim 16$), are replaced by Δ^1-pyrroline units, which function as aza bridges ($pK \sim 2$) in the inner conjugation path. Therefore the ligand becomes less basic and the oxidation potential of the central metal goes up, hence it becomes easier to remove an electron from the ligand itself [*Fuhrhop* (*70a*)].

The electronic spectrum of the MgOEP cation radical contains an unusually intense solvent- and counterion-dependent absorption band at 685 nm, which can be explained either by distortion of the macrocycle and interaction of a methine bridge with a soft counterion or by a change in the symmetry of the porphyrin orbital that contains the electron hole [*Felton* (*59*)].

The esr signal ($g = 2.0027$, linewidth $= 3.8$ G) of the MgOEP cation radical in solution accounts quantitatively for the porphyrin present [*Fuhrhop* (*78*)] and shows hyperfine splitting through the methine protons ($a_H = 1.48$ G) but not through the nitrogens [*Felton* (*59*)]. The relatively large linewidth can be explained by the degeneracy of the outer orbitals which yields some orbital moment contributions (see III), but surprisingly the chlorin esr spectra yield similar broad lines [*Fuhrhop* (*70a*)]. This phenomenon has still to be evaluated. At low temperatures (−70 °C) a dimer with an absorption band at 950 nm has been observed which did not produce an esr spectrum [*Felton* (*59*)].

Chl_a and BChl can be easily oxidized to π-cation radicals ($E_{\frac{1}{2}}^{ox} Chl_a =$ 0.5 V, $E_{\frac{1}{2}}^{ox}$ BChl $= 0.27$ V) [*Diehn* (*47*), *Fuhrhop* (*78*). *Goedheer* (*83*)]. The reaction is reversible with reducing agents, except for some irreversible side-reactions of the vinyl group [*Fuhrhop* (*78*)] and cyclopentanone ring [*Pennington* (*145*)] which usually occurs at potentials above 350 mV. The electronic spectra of these radicals are of the chlorin π-cation radical type (Fig. 1) and the esr signal consists of a single line at $g = 2.0025$. Peroxides can be decomposed in the presence of light when chlorophylls are present. The intermediacy of a Chl_a cation radical is probable [*Oster* (*135*)].

Zinc Porphyrin Cation Radicals

Zn OEP undergoes a fully reversible one-electron oxidation at 525 mV [*Fuhrhop* (*78*)] and ZnTPP at 710 mV [*Felton* (*59*)]. The visible spectra of the resulting π-cation radicals consist of one broad band covering the whole visible range. Zn OEP^+ forms a stable π–π' dimer, which produces a very strong, broad absorption band with a maximum around 930 nm [*Fuhrhop* (*79*)]. This diamagnetic dimer has exactly the same redox behavior as the monomer and is also formed when the monomeric radical is crystallized.

The esr spectrum of the $ZnOEP^+$ radical consists of one sharp band ($g = 2.0025$ linewidth $= 2.76$) with no hyperfine splittings, whereas the $ZnTPP^+$ radical ($g = 2.0025$) produces a nine-line spectrum by nitrogen splitting ($a_N = 1.43$ G). Each line is further resolved into nine lines by eight equivalent phenyl protons ($a_H = 0.24$) [*Felton* (*59*)]. These differences in the ground-state properties of ETIO-type and TPP-type porphyrin ligands can again be explained either by a simple MO argument or by the assumption of different degrees of planarity, where the less planar TPP ligand would have less degeneracy in the outer orbitals and therefore produce sharper esr signals.

A microelectrolytic cell has been constructed for the light-driven photo-oxidation of water to oxygen by NADP. The photoelectrodes were formed by many monomolecular layers of Zn TPP which were immersed in an aqueous mixture of potassium ferri- and ferrocyanide. Upon illumination, charge transfer took place across the monomolecular layers and there resulted a photoelectromotive force of approx. 1.2 V which was sufficient to oxidize water [*Wang* (*192, 193*)].

Cobalt^{III}, Nickel^{II}, Palladium^{II}, and Copper^{II} Porphyrin π-Cation Radicals

These oxidized transition-metal chelates differ from the magnesium and zinc chelates by the presence of partly occupied d orbitals, but their properties are very similar. The visible spectra resemble the monomeric zinc porphyrin π radicals, dimerization has not been reported so far, and the magnetic resonance data are fully compatible with the removal of an electron from the porphyrin π orbitals alone [*Fuhrhop* (*78*)]. Pmr shift measurements of internal standards have given clear evidence that there are two unpaired electrons in the copper porphyrin cation radicals [*Wolberg* (*212*)], but no esr spectrum is observed down to liquid nitrogen temperatures [*Fuhrhop* (*78*)] because of the line-broadening spin–spin interactions. The esr spectra of the other one-electron oxidation products clearly indicate their aromatic radical character: Ni OEP^{+} ($g = 2.0020$, linewidth 3.6 G) [*Fuhrhop* (*78*)], PdOEP (1.9991; 3.6 G) [*Fuhrhop* (*78*)] and Co^{III} OEP$^{\dot{2}+}$ [*Dolphin* (*51*)]. All give esr signals of linewidths less than 5 G and only the cobalt radical signal is split ($a_{Co} = 1.2$ G) [*Dolphin* (*51*)]. The oxidation potentials are given in Table 7.

Fe^{III} and Ru^{II} Porphyrin π-Cation Radicals

Chemical oxidation of Fe^{III} porphyrins with bromosuccinimide leads to irreversible formation of an addition product of unknown structure with a phlorin-type spectrum [*Fuhrhop* (*78*)], whereas the related electrochemical oxidation yields one-electron oxidation products, with a π-cation-radical spectrum [*Felton* (*61*)]. The latter products have been discussed in IV.6. Oxidation of Ru^{II} porphyrin carbon monoxide adducts also yields stable π-cation radicals [*Brown* (*18*)].

π Dications

Electrolysis of MgOEP$^{\dot{+}}$ at 0.9 V results in the removal of an additional electron. This diamagnetic π dication shows at uv band at 330 nm ($\varepsilon = 60{,}000$) and a broad band covering the whole visible range. MgTPP$^{\dot{+}}$, ZnTPP$^{\dot{+}}$ and CuTPP$^{\dot{+}}$ yield similar products on oxidation, and undergo easy nucleophilic addition reactions (*e.g.* with methanol), which lead to phlorin-type adducts [*Fajer* (*55*), *Felton* (*59*)].

3. π-Anion Radicals and π Dianions

Extensive work has been done on the polarographic one-electron reduction of porphyrins and their metal chelates. Usually chemical and electrochemical reductions lead to nucleophilic additions to the porphyrin macrocycle, but in carefully dried and degassed aprotic solvents the air-sensitive π mono- and di-anions can be preserved over long periods. This paragraph deals with the addition of electrons to the porphyrin, the next one will briefly summarize the vast field of reversible hydrogenations of the methine bridges, but only a brief reference will be made to the equally interesting fields of irreversible hydrogenations on the α and β positions of the pyrrole rings which lead to reductive opening of the porphyrin macrocycle [*Chapman* (*29*), *Küster* (*118*)], or to chlorin formation [*Imhoffen* (*100*)].

Chemically stable porphyrins can be reduced with sodium anthracenide [*Closs* (*35*)]:

$$e.g.\ \mathrm{Zn\,TPP} + C_{14}H_{10}\,\dot{-}\,Na^{+} \rightleftharpoons \mathrm{Zn\,TPP}\,\dot{-}\,Na^{+} + C_{14}H_{10}\,,$$

whereas porphyrins with reducible side chains (*e.g.* BChl) are better obtained by electrochemical reductions [*Clack* (*32*), *Dolphin* (*49*), *Fajer* (*55a*)]. The electronic spectra of TPP, tetraphenylchlorin, tetraphenylbacteriochlorin, deutero, and BChl anion radicals display rather sharp absorptions above 800 nm and a broad Soret band [*Dodd* (*48*), *Dolphin* (*49*), *Peychal-Heiling* (*147, 148*), *Fajer* (*55a*)]. The esr spectra usually consist of a signal at $g=2.00$ with a linewidth of about 5 G and often show hyperfine splitting [*Clack* (*32*), *Hush* (*97*)].

Diamagnetic π dianions have also been obtained by reduction in aprotic solvents and surprisingly produce an electronic spectrum which resembles the "normal" spectra of metalloporphyrins [*Closs* (*35*), *Peychal-Heiling* (*147, 148*)].

Light-driven electron transfers between Chl_a and hydroquinones [*Cho* (*30*), *White* (*200*)] as well as between Chl_a and lumiflavin [*Tu* (*184*), *185*)] have also been described and interpreted as simple models for some primary steps in photosynthesis [*Wang* (*193*)]. Through the use of β carotene as a quencher the rate constant for semiquinone radical formation in such a system was found to be 8.8×10^5 l/mol sec at $-30°$ [*Cho* (*30*)]. A descriptive theory of the relationships between the energy levels in different chlorophylls and the electron-transfer processes between them and other organic electron carriers has also been developed [*Nelson* (*134*)].

4. Reversible Hydrogen Additions to Reduced Porphyrins

Although this article deals with electron transfers with metalloporphyrins and generally does not cover the chemical reactions of the porphyrin ligand, it is felt that one exception has to be made and that the protonation of the methine bridges of porphyrin π anions should be discussed. This reaction invariably occurs when the reduction of porphyrins is carried out in protic media and is usually fully reversible with a rise in the redox potential of the solution, very much like electron addition and removal [*Closs* (*35*), *Inhoffen* (*99, 101, 102, 103*), *Kiselev* (*113*), *Mauzerall* (*129, 130*), *Peychal-Heiling* (*147, 148*), *Seely* (*156*), *Shablya* (*157*), *Shulga* (*158*), *Sidorov* (*159*), *Woodward* (*213, 214*)].

Mainly four hydrogenated porphyrin states may be expected after the protonation of the various porphyrin π anions: the porphyrinogen state (XIX) with four isolated pyrrole units (colorless, absorption bands ~200 nm), the porphomethene (XX) (orange, ~500 nm), and porphodimethene states (XXI) (purple, ~550 nm) with pyrromethene units and the bilatriene-like phlorins (XXII) (green or brownish-green, ~700–800 nm). The given structures XIX–XXII correspond to the metal chelates with divalent central ions and nonprotonated nitrogens. *Mauzerall* was the first to characterize various of such mesohydrogenated porphyrin states obtained by the photochemical reduction of porphyrins, (*129, 130, 131 a*) and *Closs* and *Closs* (*35*) demonstrated later that phlorins and porphomethenes could also be obtained by protonation of zinc TPP anions from the sodium anthracenide reduction. Similar results from electrochemical experiments on various porphyrins and chlorophyll derivatives have been thoroughly investigated and there can be no doubt that the reduction of porphyrins and metalloporphyrins followed by protonation always leads to reversible hydrogenation of the methine bridges. This pathway is followed until all four methine bridges are reduced (porphyrinogen) and then no further reaction occurs [*Küster* (*118*)]. Qualitative and quantitative electron-density arguments can explain this special reactivity of the methine carbon atoms [*Knop* (*115*)], and the theoretical prediction that in a phlorin the opposite bridge atom should be especially vulnerable to further hydrogenation has also been confirmed by various experiments on the phlorins, *e.g.* the reductive dimethylation of *Buchler* (*20*)].

XIX

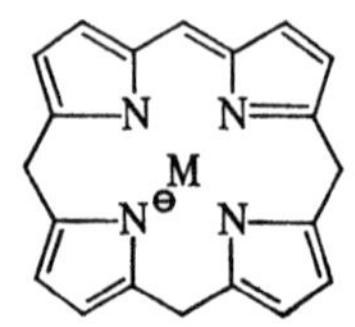

XX

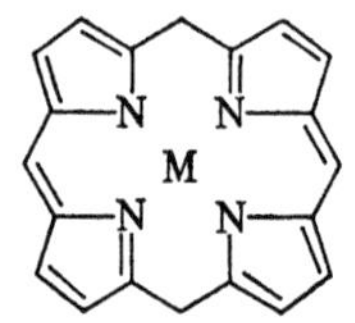

XXI

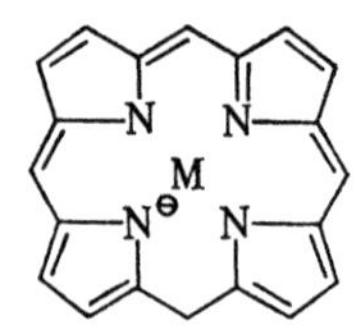

XXII

Rearrangement of metallophlorins to air-stable chlorins has been observed in various instances [*e.g.* (*Buchler* (*20*), *Closs* (*35*), *Fuhrhop* (*75*)]. The further reduction to chlorinphlorins of natural chlorophyll derivatives occurs exclusively on the β bridge, which again proves the special vulnerability to nucleophilic attack of methine bridges between pyrrole and pyrrolenine units [*Inhoffen* (*101, 102, 103*)].

The most interesting hydrogenation products of metalloporphyrins are undoubtedly the phlorins (IV, XXII), which were discovered independently by *Mauzerall* and *Woodward* [*Mauzerall* (*129, 130*), *Woodward* (*213, 214*)]. Phlorin-type compounds are easily formed under various conditions and are comparable in energy content to the completely conjugated cyclic electronic system of the porphyrin. This is evidenced by various equilibria between oxo phlorins and their porphyrin tautomers [*Bonnett* (*8*), *Fuhrhop* (*72*), *Jackson* (*106*), *Stier* (*168*)] as well as by the unusual stability of phlorins in the absence of oxygen [*Mauzerall* (*130, 131*), *Woodward* (*213, 214*)]. The physical and chemical properties of the phlorins and metallophlorins, however, differ fundamentally from those of the porphyrins: phlorins produce bile pigment-type electronic and pmr spectra with no ring-current effect, they very easily undergo irreversible addition reactions, and they are much easier to oxidize [*e.g.* $E_{\frac{1}{2}}^{Ox}$(α-Oxo-OEP): -50 mV] [*Fuhrhop* (*73*)]. If the porphyrins are taken as analogs to the 16-annulene dianion, then the phlorins behave like the related open-chain polyenes. It is this fundamental difference combined with the easy convertibility of porphyrins into phlorins and vice versa that makes systematic comparative studies on these metal chelates so attractive.

5. Summary

The reversible porphyrin ligand reductions and oxidations are governed by the activation or deactivation of the π-electron conjugated system through the electrostatic action of the framework containing the more or less polar metal–nitrogen bands. A simple rationalization of this finding has been proposed with the introduction of an "inductive parameter" for various metal ions in the porphyrin cavity [*Fuhrhop* (*74*)]. That inductive effects of the central ions predominate over conjugative effects from π–$d_{(xz,yz)}$ interactions has also been concluded from theoretical considerations [*Gouterman* (*85, 86*)], and the calculated net charges of the porphyrin ring in various metal complexes [*Zerner* (*219*)] correlate with the potentials of reversible one-electron reactions of the porphyrin ligand in qualitatively the same way as the induction parameters of the metal ions.

The behavior of the porphyrin ligand in irreversible chemical reactions follows the same trends, as exemplified by the two extreme cases of the magnesium porphyrins, where the ligand π system is extraordinarily nucleophilic, and tin^{IV} porphyrins, where it reacts as an electrophile.

The typical reactions of the porphyrin ligand can be compared to those of aromatic compounds, whereas interruption of the macrocyclic conjugation path at one methine bridge leads to a more polyene-type reactivity pattern of the resulting phlorins.

VI. Redox Reactions of Metalloporphyrins in Biological Systems

1. Cytochromes

The concept of the photosynthetic and respiratory systems is that they are based on the rapid flow of electrons through a graded sequence of redox steps mediated by a continuous sequence of carriers [*Change* (*27*)]. Hemoproteins play an important role here, as do other "organic" carriers like quinones, flavins, and copper and iron proteins.

Although the substitution patterns of the porphyrins occurring in the different cytochromes are known [*Falk* (*56*)], no correlation between midpoint potentials and chemical structure has been found so far, because with most cytochromes the axial ligands and their positions are not known. The rate of oxidation of the iron in isolated cytochrome c is very fast [*Sutin* (*171*)]. Direct measurements of redox potentials with hemoproteins have been discussed recently [*Betso* (*5*)]. An anaerobic potentiometric titration of leghemoglobin with anthrahydroquinone sulphonate yielded a midpoint potential of -20 mV [*Henderson* (*92*)] but the importance of this hemoprotein, which is always present in active nitrogen-fixing root nodules, is not known. A detailed description of a potentiometric titration of a hemoprotein with the aid of mediators has been given for horseradish peroxidase [*Harbury* (*89*)]. Fe^{II} cytochromes also reduce alkylhalogenides [*Wade* (*188*)], as has been described for the Fe^{II} porphyrins (see IV.6).

A hypothesis that, for the electron transfer through the protein fabric of hemoproteins, tyrosine and tryptophane radicals are involved in the protein case has been discussed [*Winfield* (*209*)]. However, the conditions necessary [*King* (*112*)], to obtain such protein radicals (uv light, 1% H_2O_2; $IrCl_6$) are unphysiological and, even if they could be formed in vivo, it is hard to see how irreversible side-reactions of these short-lived radicals could be prevented. The complete reversibility of such unphysiological oxidations has not been demonstrated. Another hypothesis proposes that electron transfer is primarily over exposed heme edges, especially when the iron is blocked by the protein matrix [*Change* (*28*)]. Evidence has been produced that, during a redox cycle in which the iron is reduced and reoxidized, the methine protons exchange with the solvent [*Castro* (*24*)]. From this it has been postulated that the primary peripheral attack is on the porphyrin ring with the formation of phlorin radicals in the reduction of Fe^{III} hemoproteins. More conclusive experiments are needed to prove or disprove any of the above models.

Anaerobic titrations coupled with spectrophotometric measurements in the Soret-band region of mitochondria and the activation by pulsed light of selectively poised photosynthetic systems have clearly shown that cytochromes are responsible for a number of redox reactions in the range -250 to $+150$ mV in the corresponding electron transfer chains [*Wilson* (*207, 208*)]. The central switch mechanism which determines the actual rate of electron transfer in the respiratory chain has been provisionally located at two cytochrome sites (a_3, b_1), where a high-potential redox couple is thought to be in quasi-equilibrium with a low-potential form (*e.g.* $-270/0$ mV for the cyt $b_{T\,low}/b_{T\,high}$ conversion) and where this transformation is coupled with ATP formation. An electron transfer from or to the cyt $b_{T\,high}$ would then be possible only with the electron carrier system on the oxygen side, whereas ATP formation or a similar energy-absorbing reaction at this site would produce the low-potential cyt $b_{T\,low}$ form, which then could only react with the NADH side of the chain [*Chance* (*27*)]. It has been observed that addition of ATP can reverse the electron flow in the respiratory system. This is considered to arise from the effect of the system ATP/ADP, P_i, H^+ on the high- and low-potential forms of cyt b_T [*Chance* (*27*), *Devault* (*46*)]. This recent discovery, that addition of ATP can lower or raise the apparent midpoint potential of biological redox couples, may ultimately prove to be the key to an understanding of the energy transduction from electron transport chains to high-energy phosphate bondings.

2. Heme Proteins which Bind and/or Activate Molecular Oxygen

Hemoglobin

The oxygen adduct of hemoglobin (HbO_2) is diamagnetic and shows a "normal" metalloporphyrin electronic spectrum. It was mainly these two experimental facts which led to the first-postulated HbO_2 structure where an oxygen molecule is bound covalently to a low-spin Fe(II) ion, the O—O bond being at an angle either close to 45° [*Pauling* (*141*, *142*)] or parallel [*Griffith* (*87*)] to the porphyrin plane. Both these facts, however, can be interpreted differently [*Wittenberg* (*210*)]. The electronic spectra of HbO_2 and low-spin alkaline ferric porphyrins are essentially the same, and it has been postulated that two spins on adjacent atoms can combine so as to cancel their paramagnetism [*Peisach* (*144*), *Wittenberg* (*210*)]. Such an interpretation, together with some analogies drawn with cobaltic porphyrin oxygen adducts, would favor a ferric low-spin iron to which a superoxide anion is bound. Results from Fe Kβ fluorescence spectra of HbO_2 have been interpreted accordingly [*Koster* (*117a*)]. By this technique it has been shown that the iron in HbO_2 possesses one unpaired electron, explained as a d^2sp^3 hybridization of a trivalent iron ion, as in the complex cyanides. If a charge of $+3$ is found on the iron, the O_2 ligand is effectively the superoxide anion. Early chemical evidence for this new hypothesis for HbO_2 came from an experiment in which the oxygen that was released after the acidic denaturation of the globin was shown to be chemical activated oxygen. It reacted rapidly with ascorbic acid and other organic substrates [*Lemberg* (*120*, *121*)]. None of the evidence for the superoxide anion nature of the oxygen in HbO_2, however, can be taken as conclusive.

The structural changes in the hemoglobin subunits, accompanying the oxygenation of hemoglobin have been attributed to a 0.8-Å shift of the iron atoms out of the porphyrin planes [*Perutz* (*145*)]. If this somewhat controversial viewpoint, which is based on a rather intuitive interpretation of X-ray and magnetic data, proves to be correct it would be an outstanding example of an allosteric rearrangement in which molecular cause and effect had been clarified, it could thus become one of the cornerstones of mechanistic enzyme chemistry.

Dioxygenases

The heme enzyme, tryptophan dioxygenase, activates molecular oxygen to undergo the following reaction [*Cragh* (*125*), *Yamamoto* (*215*), *Marks* (*128*)]:

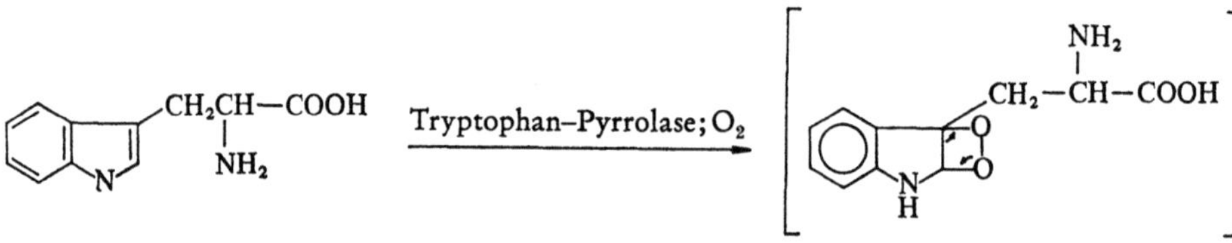

XXIII XXIV

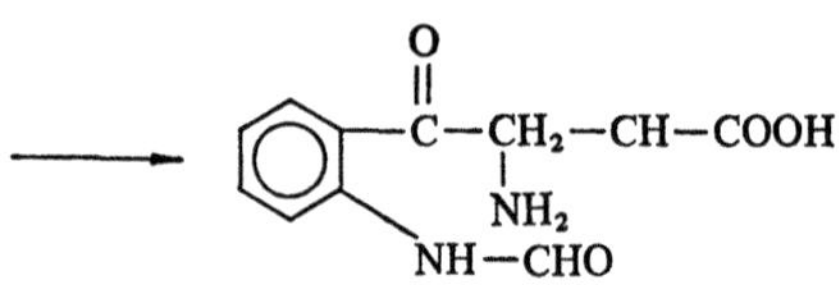

XXV

In this case the formation of the superoxide anion from molecular oxygen has been substantiated by quenching experiments with superoxide dismutase [*Brady* (*14*), *Strobel* (*169*), *Hirata* (*93*)], although these cannot be accepted as definite proof because the specifity of the enzyme for superoxide anions has not been conclusively demonstrated.

This reaction of the oxygen molecule activated by a heme enzyme is somewhat reminiscent of the photooxygenation of metallo-porphyrins mentioned in V.1 (Formulas XII–XV).

Hydroxylating Enzymes

A microsomal cytochrome, called "cytochrome P-450", catalyses substrate hydroxylations by molecular oxygen. The two-electron process also involves the formation of an oxygen intermediate where the oxygen may be in the "oxenoid" form and the iron probably in the Fe^{II} oxidation state [*Ullrich* (*186*)].

$$Fe^{2+} + O_2 \rightleftharpoons \underset{\text{"superoxide" complex}}{[Fe^{3+}O_2^-]^{2+}} \overset{+e}{\rightleftharpoons} \underset{\text{"oxenoid" complex}}{[Fe{-}O{-}O]^+}$$

Such highly negatively charged oxygen species with two electrons in antibonding orbitals could explain the high reactivity of P-450-activated oxygen and the need for low potential reductants in these enzymatic reactions, and it is in accord with the extremely low reduction potential of approx. −600 mV of "cytochrome P-450" [*Waterman* (*196*)]. The redox potential of this hemoprotein was determined by setting equilibrium potentials anaerobically with large excesses of redox indicators and determining the relative concentration of the oxidized cytochrome P-450 by esr. Typical biological substrates for enzymatic hydroxylations which may involve P-450 are heme [*Tenhunen* (*173*)], steroids [*Wilson* (*206*)] (*e.g.* XXVI–XXVII):

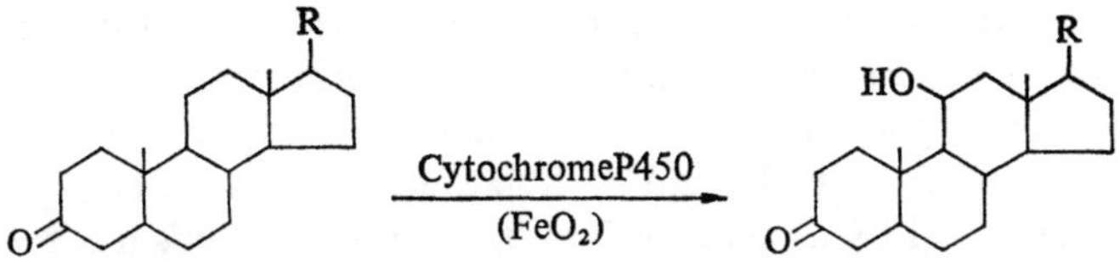

XXVI XXVII

and camphor [*Ishimura* (*104a*)]. It has been proposed that, during hydroxylation with this enzyme, low-spin Fe^{III} cytochrome-P-450 becomes high-spin upon complex formation with substrate, and is then reduced to high-spin reduced Fe^{II} cytochrome-P-450 [*Tsai* (*177*)]. Reducing power is provided by microsomal electron transport chains originating in DPNH or TPNH [*Ullrich* (*186*)].

3. Catalases and Peroxidases

The biological importance of these hemoproteins is not clear, although many in-vitro experiments have shown that catalase decomposes H_2O_2 extremely rapidly and may therefore serve as a natural safeguard against peroxide poisoning. Also it has been shown that both peroxidases and catalases oxidize a variety of substrates in the presence of oxygen or hydrogen peroxide [*Brill* (*15*)].

When free peroxidases or catalase react with hydroperoxides, stable green substances, called compounds I and II, are formed; they have absorption spectra of π-cation-radical type [*Blumberg* (*7*), *Brill* (*16*)]. This has been taken as evidence that stable Fe^{III} porphyrin cation radicals are formed by peroxide oxidation in these particular heme proteins [*Borg* (*10*), *Dolphin* (*51*)]. It is not known why the hydrogen peroxide does not attack the porphyrin ring, as is normal with hemoproteins, nor is it certain whether this radical formation has any biological implication.

4. Chlorophyll π-cation Radicals in Photosynthesis

Green plants [*Commoner* (*37a*)], photosynthetic bacteria [*Sogo* (*163*)], and isolated parts of the photosynthetic redox chain [*McElroy* (*132*)] produce an intense esr signal in presence of light, decaying rapidly in the dark. This is called signal I in green plants and signal B I in bacteria, as opposed to signals II and B II, which have a slower decay time and are not as intense. Signals I and B I are also induced at 77 °K. The action spectrum of radical formation follows that of active-center chlorophyll (P-700) rather than of bulk chlorophyll and the cross-section for photocapture for paramagnetism in spinach chloroplasts was about 150 chlorophyll molecules, i.e. about the size of the photosynthetic unit. These and other data have been thoroughly reviewed recently [*Feher* (*58*), *Kohl* (*116*), *Weaver* (*198*)] and clearly indicate that this esr signal arises from a species directly involved in the primary events of photosynthesis. The signals I and B I are about 7.5 and 9.7 G broad, have g values of 2.0025 ± 0.0002, and no hyperfine structure. The signals are however narrowed by a factor of 2 to 3, when the organism is grown in D_2O, which indicates a free radical containing hydrogen atoms [*Kohl* (*117*), *McElroy* (*132*)]. From all that is known about photosynthesis it is quite clear that chlorophyll π-cation radicals are the primary candidates for this light-induced esr signal. The similarity of the midpoint reduction potential of this species, the oxidation potential of a chemical oxidation leading to a similar esr signal and the potentials of the bleaching of P-700 also supply strong evidence for the correctness of this assumption [*Beinert* (*4*), *Fuhrhop* (*78*)]. Direct quantitative experiments on the identity of the spectrophotometric P-700 concentration, and the number of unpaired electrons, however, yielded a ratio of 1 to 4, with the data clustering around 2, indicating that other chlorophyll molecules than those of the active center contribute to the esr signal [*Kohl* (*116*)]. The significance of these additional radicals is not yet known. The similarity of the rise and decay kinetics of the esr signal and of the P-700 spectrum (or P-865 or P-800 in bacteria) make it quite clear that the observed processes are identical and can be attributed to the formation of a π-cation radical of an active chlorophyll entity, which in a dark reaction rapidly takes up an electron from the coordinated partner in the photosynthetic redox chain. In bacteria and algae this is probably a low-potential iron protein [*Feher* (*58*)], *McElroy* (*132*), *Ke* (*110a,b*)].

The primary conversion of light energy into chemical redox energy in photosynthesis can therefore be described by a simple scheme [*Kok* (*111*)]:

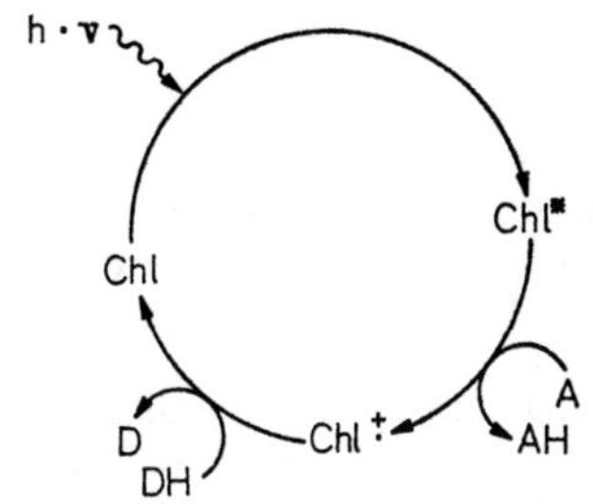

Fig. 5. Simplified scheme of the conversion of light energy into redox energy in photosynthesis (KOK 111).

The biochemical conversion of magnesium protoporphyrin monomethylester to protochlorophyll probably also involves the primary formation of a π-cation radical [*Cox* (*41*)]. Whether the hydrogenation of protochlorophyll to chlorophyll involves phlorin intermediates [*Woodward* (*214*)] is still not known.

VII. Conclusion

The redox chemistry of metalloporphyrins is dominated by strong σ-electron donation from the nitrogens to the metal ion. Metals of low electronegativity and low oxidation numbers activate the porphyrin ligand with respect to oxidation and deactivate it with respect to reductions. The opposite is true for metals of high electronegativity and/or oxidation number. Such reversible reactions of the macrocyclic ring are of importance in biological systems containing magnesium porphyrins. On the other hand, electron-rich transition-metal ions with formally high oxidation numbers are stable in the porphyrin cavity and tend to transfer some electrons to axial ligands. This effect is probably of biological importance in the hemoproteins, which function in electron transfers and oxygen activation.

What is most needed now, in the opinion of the author, is a detailed model rationalizing the interactions between the in-plane metal–nitrogen bonds and the π and d orbitals of the porphyrins and axial ligands, as well as the d orbitals of the metals that do not lie in this plane. Such an insight should account for the main influence of the three partners in the porphyrin metal complexes on each other; many more exciting experiments could then be planned.

Acknowledgements. I thank Doz. *H. Brockmann*, Prof. *P. Hemmerich*, Prof. *H. H. Inhoffen*, Prof. *D. Mauzerall*, Dr. *J. Subramanian* and Prof. *D. Whitten* for carefully reading the manuscript and for their many valuable suggestions. The Deutsche Forschungsgemeinschaft supported this work with a grant and a Habilitanden Fellowship.

VIII. References

1. *Adler, A. D.:* In: Hemes and Hemoproteins. (*Chance, B., Estabrook, R. W., Yonetani, T.*). New York: Academic Press 1966.
2. *Alben, J. O., Fuchsmann, W. H., Beaudreau, C. A., Caughey, W. S.:* Biochem. *7*, 624, (1968).
3. *Barron, E. S. G.:* J. Biol. Chem. *121*, 285 (1937).
4. *Beinert, H., Kok, B., Hoch, G.:* Biophys. Biol. Res. Comm. *7*, 209 (1962).
5. *Betso, S. R., Klapper, M. H., Anderson, L. B.:* J. Am. Chem. Soc. *94*, 8197 (1972).
6. *Blumberg, W. E., Peisach, J.:* J. Biol. Chem. *240*, 870 (1965).
7. — — *Wittenberg, B. A., Wittenberg, J. B.:* J. Biol. Chem. *243*, 1854 (1968).
8. *Bonnett, R.:* J. Chem. Soc. Perkin I *1972*, 2540.

8a. *Bonnett, J. J., Eaton, S. S., Eaton, G. R., Holm, R. H., Ibers, J. A :* J. Am. Chem. Soc. *95*, 2141 (1973).

9. *Bonnett, R., Brewer, R.:* Tetrahedron Letters *1970*, 2579.
10. *Borg, D. C.:* In: Biological Applications of ESR, p. 165. (*Swartz, H. M., Bolton, J. R., Borg, D. C.*, Eds.) New York: Wiley 1972.
11. *Boucher, L. J.:* Coord. Chem. Rev. *7*, 289 (1972).
12. — *Garber, H. K.:* Inorg. Chem. *9*, 2644 (1970).
13. *Boylan, D. B., Calvin, M.:* J. Am. Chem. Soc. *89*, 5472 (1967).
14. *Brady, F. O., Forman, H. J., Feigelson, P.:* J. Biol. Chem. *246*, 7119 (1971).
15. *Brill, A. S.:* Comprehensive Biochem. *14*, 447 (1966).
16. — *Williams, R. J. P.:* Biochem. J. *78*, 253 (1961).
17. *Brody, M., Broyde, S. B., Yeh, C. C., Brody, S. S.:* Biochem. *7*, 3007 (1968).
18. *Brown, G. M., Hopf, F. R., Ferguson, J. A., Meyer, T. J., Whitten, D. G.:* (in preparation).
19. *Buchler, J. W., Eikelman, G., Puppe, L., Rohbock, K., Schneehage, H.H., Weck, D.:* Ann. Chem. *745*, 135 (1971).
20. — *Puppe, L.:* Ann. Chem. *740*, 142 (1970).
21. — *Rohbock, K.:* Inorgan. Nucl. Chem. Lett. *8*, 1073 (1972).
22. *Calvin, M.:* Rev. Pure. Appl. Chem. *15*, 1 (1965).
23. — *Loach, P. A., Yamamoto, A.:* In: Theory and Structure of Complex Compounds, p. 13 (*Jezovska-Trzebistowska, B.*, ed.). Macmillan 1964.
24. *Castro, C. E., Davis, H. F.:* J. Amer. Chem. Soc. *91*, 5405 (1969).
25. *Caughey, W. S., Fujimoto, W. Y., Johnson, B. P.:* Biochem. *5*, 3830 (1966).
26. *Cavaleiro, I. A. S., Smith, K. M.:* Chem. Comm. *1971*, 1384.
27. *Chance, B.:* FEBS Lett. *23*, 3 (1972).
28. — *Lee, C., Mela, L., de Vault, D.:* In: Structure and Function of Cytochromes, p. 475. University Tokyo Press 1968.
29. *Chapman, R. A., Roomi, M. W., Morton, T. C., Krajcarski, D. T., Mac Donald S. F.:* Can. J. Chem. *49*, 3544 (1971).
30. *Cho, D. R., Tollin, G.:* Photochem. Photobiol. *8*, 317 (1968).
31. *Chow, B. C., Cohen, A.:* Bioinorg. Chem. *1*, *57* (1971).
32. *Clack, D. W., Hush, N. S.:* J. Amer. Chem. Soc. *87*, 4238 (1965).
33. *Clark, W. M.:* Oxidation Potentials of Organic Systems, Williams and Wilkins (1960).
34. *Clarke, D. A., Dolphin, D., Grigg, R., Johnson, A. W., Pinnock, H. A.:* J. Chem. Soc. (C), *1968*, 881.
35. *Closs, G. L., Closs, L. E.:* J. Amer. Chem. Soc. *85*, 818 (1963).

35a. *Cohen, I. A., Ostfeld, D., Lichtenstein, B.:* J. Am. Chem. Soc. *94*, 4522 (1972).

36. — *Coughey, W. S.:* Biochem. *7*, 636 (1968).
37. — *Jung, C., Governo, T.:* J. Amer. Chem. Soc. *94*, 3003 (1972).
37a. *Commoner, B., Townsend, J., Pake, G.:* Nature *174*, 689 (1954).
38. *Corwin, A. H., Bruck, S. D.:* J. Am. Chem. Soc. *80*, 4736 (1958).
39. — *Erdman, J.:* J. Am. Chem. Soc. *68*, 2473 (1946).
40. — *Reyes, Z.:* J. Am. Chem. Soc. *78*, 2437 (1956).
41. *Cox, M. T., Howarth, T. T., Jackson, A. H., Kenner, G. W.:* J. Am. Chem. Soc. *91*, 1232 (1969).
42. *Cullen, D., Meyer, E., Srivastava, T. S., Tsutsui, M.:* Chem. Comm. *1972*, 584.
43. — — — — J. Am. Chem. Soc. *94*, 7603 (1972).
44. *Davis, D. G., Martin, R. F.:* J. Am. Chem. Soc. *88*, 1365 (1966).
45. — *Montalvo, J. G.:* Analyt. Chem. *41*, 1195 (1969).
46. *Devault, D.:* Biochim. Biophys. Acta *226*, 193 (1971).
47. *Diehn, B., Seely, G. R.:* Biochim. Biophys. Acta *153*, 862 (1968).
48. *Dodd, J. W., Hush, N. S.:* J. Chem. Soc. *1964*, 4607.
49. *Dolphin, D., Felton, R. H.:* Accts. Chem. Res. (in press).
50. — — *Borg, D. C., Fajer, J.:* J. Am. Chem. Soc. *92*, 743 (1970).
51. — *Forman, A., Borg, D. C., Fajer, J., Felton, R. H.:* Proc. Natl. Acad. Sci. *68*, 614 (1971).
52. *Edwards, L., Dolphin, D. H., Gouterman, M.:* J. Mol. Spec. *35*, 90 (1970).
53. *Elvidge, J. A., Lever, A. B. P.:* Proc. Chem. Soc. *1959*, 195.
54. *Engelsma, G., Yamamoto, A., Markham, E., Calvin, M.:* J. Phys. Chem. *66*, 2517 (1962).
55. *Fajer, J., Borg, D. C., Forman, A., Dolphin, D., Felton, R. H.:* J. Am. Chem. Soc. *92*, 3451 (1970).
55a. — — — — — J. Am. Chem. Soc. *95*, 2739 (1973).
56. *Falk, J. E.:* Porphyrins and Metalloporphyrins, Elsevier, *1964*.
57. — *Perrin, D. D.:* In: Haematin Enzymes, p. 56 (*Falk, J. E., Lemberg, R., Morton, R. K.* Eds). Pergamon 1961.
58. *Feher, G.:* EPR with Applications to Selected Problems in Biology, Gordon and Breach Publ., N.Y. 1970, Chapter III.
59. *Felton, R. H., Dolphin, D., Borg, D. C., Fajer, J.:* J. Am. Chem. Soc. *91*, 196 (1969).
60. —*Linschitz, H.:* J. Am. Chem. Soc. *88*, 1113 (1966).
61. — *Owen, G. S., Dolphin, D., Fajer, J.:* J. Am. Chem. Soc. *93*, 6332 (1971).
62. *Ferguson, J. A., Meyer, T. J., Whitten, D. G.:* Inorg. Chem. *11*, 2766 (1971).
63. *Fischer, H., Herrle, K.:* Hoppe-Seylers Z. physiol. Chem. *251*, 85 (1938).
64. — *Treibs, A., Zeile, K.:* Hoppe-Seylers Z. physiol. Chem. *195*, 1 (1930).
65. *Fleischer, E. B.:* Accts. Chem. Res. *3*, 105 (1970).
66. — *Laszlo, A.:* Inorg. Nucl. Chem. Lett. *5*, 373 (1969).
67. — *Srivastava, T.:* Inorg. Chim. Acta *5*, 151 (1971).
68. — — J. Am. Chem. Soc. *92*, 5518 (1970).
69. — *Thorps, R., Venerable, D.:* Chem. Comm. *1969*, 475.
70. *Fuhrhop, J.-H.:* (unpublished work).
70a. — Z. Naturforsch. *25b*, 255 (1970).
70b. — Angew. Chem. (in press).
71. — Tetr. Lett. *1969*, 3205.
72. — Chem. Comm. *1970*, 781.
73. — *Besecke, S., Subramanian, J.:* Chem. Comm. *1973*, 1.
74. — *Kadish, K., Davis, D. G.:* J. Am. Chem. Soc. *95*, 5140 (1973).
75. — *Lumbantobing, T.:* Tetr. Lett. *1970*, 2815.
76. — *Mauzerall, D.:* J. Am. Chem. Soc. *91*, 4174 (1969).

77. — — Photochem. Photobiol. *13*, 453 (1971).
78. — — J. Am. Chem. Soc. *90*, 3875 (1968).
79. — *Wasser, P. K. W., Riesner, D., Mauzerall, D.:* J. Am. Chem. Soc. *94*, 7996 (1972).
80. *George, P.:* Biochem. J. *55*, 220 (1953).
80a. — *Hanania, G. I. H., Eaton, W. A.:* Ref. 1, p. 267.
81. — *Ingram, D. J. E., Bennett, J. E.:* J. Am. Chem. Soc. *79*, 1870 (1957).
82. *Gibson, J. F., Ingram, D. J. E.:* Nature *178*, 871 (1956).
83. *Goedheer, J. C.:* In: The Chlorophylls, p. 164ff. (*Vernon, L. P., Seely, G. R.*, Eds.). New York: Academic Press 1966.
84. — *de Haas, G. H. H., Schuller, P.:* Biochim. Biophys. Acta *28*, 278 (1958).
85. *Gouterman, M.:* J. Chem. Phys. *30*, 1139 (1959).
86. — *Zerner, M., Kobayashi, G.:* Theoret. Chim. Acta *6*, 363 (1966).
87. *Griffith, J. S.:* Proc. Roy. Soc. *A 235*, 23 (1956).
88. *Hambright, P.:* Coord. Chem. Rev. *6*, 247 (1971).
89. *Harbury, H. A.:* J. Biol. Chem. *225*, 1009 (1957).
90. *Havemann, R., Haberditzl, W., Mader, K. H.:* Z. Phys. Chem. *218*, 71 (1961).
91. *Heilbronner, E., Bock, H.:* Das HMO-Modell und seine Anwendung, p. 168ff. Verlag Chemie 1968.
92. *Henderson, R. W., Appleby, C. A.:* Biochim. Biophys. Acta *283*, 187 (1972).
92a. *Hill, H. A. O., Norgett, M. M.:* J. Chem. Soc. *1964*, 1476.
93. *Hirata, F., Hayaishi, O.:* J. Biol. Chem. *246*, 7825 (1971).
94. *Hoard, J. L.:* Science *174*, 1295 (1971).
95. *Hoijtink, G. J.:* Rec. trav. chim. *77*, 555 (1958).
96. *Horio, T., Kamen, M. D.:* Biochim. Biophys. Acta *48*, 266 (1961).
97. *Hush, N. S., Rowlands, J. R.:* J. Am. Chem. Soc. *89*, 2976 (1967).
98. — *Dodd, J. W.:* J. Chem. Soc. *1964*, 4607.
99. *Inhoffen, H. H.:* Naturwissenschaften *55*, 457 (1968).
100. — *Buchler, J. W., Jäger, P.:* Progr. Chem. Organ. Nat. Prod. XXVI, 284 (1968).
101. — *Jäger, P., Mählhop, K., Mengler, C. D :* Ann. Chem. *704*, 188 (1967).
102. — — Tetrahedron Lett. *21*, 1317 (1964).
103. — *Mählhop, R.:* Tetrahedron Lett. 1966, 4283.
104. *Tizuka, T., Kotani, M., Yonetani, T.:* Biochim. Biophys. Acta *167*, 257 (1968).
104a. *Ishimura, Y., Ullrich, V., Peterson, J. A.:* Biochem. Biophys. Acta *267*, 363 (1972).
105. *Jackman, L. M., Sternhell, S.:* Applications of NMR Spectroscopy in Organic Chemistry, p. 96ff. Oxford: Pergamon Press 1969.
106. *Jackson, A. H., Kenner, G. W., Smith, K. M.:* J. Chem. Soc. (C) *1968*, 302.
107. *James, B. R., Stynes, D. V.:* J. Am. Chem. Soc. *94*, 6225 (1972).
108. *Kadish, K., Davis, D. G.:* Ann. N. Y. Acad. Sci. *206*, 495 (1973).
109. — — *Fuhrhop, J.-H.:* Angew. Chem. *84*, 1072 (1972).
110. *Kane, A. R., Yalman, R. G., Kenney, M. E.:* Inorg. Chem. *7*, 2588 (1968).
110a. *Ke, B.:* Arch. Biochem. Biophys. *152*, 70 (1972).
110b. — *Beinert, H.:* Biochim. Biophys. Acta *305*, 689 (1973).
111. *Kok, B.:* In: Plant Biochemistry (*Bonner, J., Varner, J. E.*, Eds.). New York: Academic Press 1965.
112. *King, N. K., Looney, F. D., Winfield, M. E.:* Biochim. Biophys. Acta *133*, 65 (1967).
113. *Kiselev, B. A., Kozlov, Y. N., Yevstigneev, V. B.:* Biofiz. *15*, 594 (1970).
113a. *Kivelson, D., Lee, S. K.:* J. Chem. Phys. *41*, 1896 (1964).

114. *Kneubühl, F. K., Koski, W. S., Caughey, W. S.:* J. Am. Chem. Soc. *83*, 1607 (1961).
115. *Knop, J. V., Fuhrhop, J.-H.:* Z. Naturforsch. *25b*, 729 (1970).
116. *Kohl, D. H.:* In: Biological Applications of ESR, p. 213. (*Swartz, H. M., Bolton, I. R., Borg, D. C.*, Eds.). New York: Wiley 1972.
117. *Kohl, D., Townsend, J., Commoner, B., Crespi, H., Katz, J.:* Nature *206*, 1105 (1965).
117a. *Koster, A. S.:* J. Chem. Phys. *56*, 3161 (1972).
118. *Küster, W.:* Hoppe-Seylers Z. physiolog. Chem. *82*, 467 (1912).
118a. *LaMar, G. N., Walker, F. A.:* J. Am. Chem. Soc. *95*, 1783 (1973).
119. *Lang, G.:* Quart. Rev. Biophys. *3*, 1 (1970) .
120. *Lemberg, R., Legge, J. W.:* J. Proc. Roy. Soc. N. S. Wales *72*, 62 (1938).
121. — — Haematin Compounds and Bile Pigments, Interscience 1949.
122. *Lever, A. B. P.:* In: Adv. Inorgan. Chem. Radiochem., p. 28ff. (*Emeleus, H. J., Sharpe, A. G.*, Eds.). New York: Academic Press 1965.
123. *Lexa, D., Lhoste, J. M.:* Experientia Suppl. *18*, 395 (1971).
124. *Loach, P. A., Calvin, M.:* Biochem. *2*, 361 (1963).
125. *MacCragh, A., Koski, W. S.:* J. Am. Chem. Soc. *87*, 2496 (1965).
126. *Malley, M., Feher, G., Mauzerall, D.:* J. Mol. Spec. *26*, 320 (1968).
127. *Maricondi, C., Swift, W., Straub, D. K.:* J. Am. Chem. Soc. *91*, 5205 (1969).
128. *Marks, G. S.:* Heme and Chlorophyll, p. 6ff. van Nostrand 1969.
129. *Mauzerall, D.:* J. Am. Chem. Soc. *82*, 1832 (1960).
130. — J. Am. Chem. Soc. *84*, 2437 (1962).
131. — *Feher, G.:* Biochim. Biophys. Acta *88*, 658 (1964).
131a. — *Granick, S.:* J. Biol. Chem. 232, 1141 (1958).
132. *McElroy, J. P., Feher, G., Mauzerall, D. C.:* Biochim. Biophys. Acta *267*, 363 (1972).
132a. *McEwen, W. K.:* J. Am. Chem. Soc. *58*, 1124 (1936).
133. *McLees, B. D., Caughey, W. S.:* Biochemistry 7, 642 (1968).
134. *Nelson, R. C.:* Photochem. Photobiol. *8*, 441 (1968).
134a. *Ogoshi, H., Watanabe, E., Yoshida, Z., Kincaid, I., Nakamoto, K.:* J. Am. Chem. Soc. *95*, 2845 (1973).
134b. — *Omura, T., Yoshida, Z.:* J. Am. Chem. Soc. *95*, 1666 (1973).
135. *Oster, G., Lu, R.:* In: Currents in Photosynthesis, p. 1. (*Thomas, J. B., Goedheer, J. C.*, Eds.). Rotterdam: Donker 1966.
136. *Ostfeld, D., Tsutsui, M., Hrung, C. P., Conway, D. C.:* J. Coord. Chem. *2*, 101 (1972).
137. *Oth, J. F. M., Baumann, H., Gilles, J. M., Schröder, G.:* J. Am. Chem. Soc. *94*, 3498 (1972).
138. *Park, R. B., Pon, N. G.:* J. Mol. Biol. *6*, 105 (1963).
139. *Parnemann, H.:* Thesis, Braunschweig 1964.
140. *Paul, K.-G., Theorell, H., Åkeson, Å.:* Acta Chem. Scand. 7, 1284 (1953).
141. *Pauling, L., Coryell, C. D.:* Proc. Natl. Acad. Sci. USA, *22*, 210 (1936).
142. — In: Hemoglobin, p. 57 (*Roughton, F. W. J., Hendrews, J. C.*, eds.). Butterworths 1949.
143. *Pearson, R. G.:* J. Am. Chem. Soc. *85*, 3533 (1963).
144. *Peisach, J., Blumberg, W. E., Wittenberg, B. A., Wittenberg, J. A.:* J. Biol. Chem. *243*, 1871 (1968).
145. *Pennington, F. C., Strain, H. H., Svec, W. A., Katz, J. J.:* J. Am. Chem. Soc. *89*, 3875 (1967).
146. *Perutz, M. F.:* Nature *228*, 726 (1970).
147. *Peychal-Heiling, G., Wilson, G. S.:* Analyt. Chem. *43*, 545 (1971).

148. — — Analyt. Chem. *43*, 550 (1971).
149. *Phillips, J. N.:* Rev. Pure Appl. Chem. *10*, 35 (1960).
149a. *Pullman, B., Pullman, A.:* Quantum Biochemistry, p. 423. New York: Interscience, 1963.
150. *Rabinowitch, E., Weiss, J.:* Proc. Roy. Soc. *A162*, 251 (1937).
150a. *Richards, P. L., Caughey, W. S., Eberspaecher, H., Feher, G., Malley, M.:* J. Chem. Phys. *47*, 1187 (1967).
151. *Roberts, E. M., Koski, W. S.:* J. Am. Chem. Soc. *82*, 3006 (1960).
152. — — *Caughey, W. S.:* J. Chem. Phys. *34*, 591 (1961).
152a. *Rohbock, K.*: Thesis, Aachen 1972.
153. *Rothemund, P., Menotti, A. R.:* J. Am. Chem. Soc. *70*, 1808 (1948).
154. *Sadavisan, N., Fleischer, E. B.:* J. Inorg. Nucl. Chem. *30*, 591 (1968).
155. *Salek, A.:* Diplomarbeit 1970. Braunschweig.
155a. *Schrauzer, G. N.:* Accts. Chem. Res. *1*, 97 (1968).
156. *Seely, G. R., Talenadge, K.:* Photochem. Photobiol. *1964*, 195.
157. *Shablya, A. V., Terenin, A. N.:* Optika i Spektr. *9*, 280 (1960) (Engl. Transl.).
158. *Shulga, A. M., Gurinovich, G. P., Sevchenko, A. N.:* Dokl. Akad. Nauk SSSR *169*, 1206 (1966).
159. *Sidorov, A. N.:* In: Elementary Photoprocesses in Molecules (*Neoporent, B. S.*, ed.) Plenum Press 1968.
160. *Simpson, W. T.:* J. Chem. Phys. *17*, 1218 (1949).
161. *Smith, K. M.:* Quart. Rev. *25*, 31 (1971).
162. — Chem. Comm. *1971*, 540.
163. *Sogo, P., Pon, N., Calvin, M.:* Proc. Natl. Acad. Sci. *43*, 387 (1957).
164. *Sovocool, G. W., Hopf, F. R., Whitten, D. G.:* J. Am. Chem. Soc. *94*, 4350 (1972).
165. *Stanienda, A.:* Z. Phys. Chem. Neue Folge *52*, 254 (1967).
166. — Z. Phys. Chem. *229*, 259 (1964).
167. — Naturwissenschaften *52*, 105 (1965).
168. *Stier, E.:* Z. Physiolog. Chem. *272*, 239 (1942).
169. *Strobel, H. W., Coon, M. J.:* J. Biol. Chem. *246*, 7826 (1971).
170. *Stynes, H. C., Ibers, J A.:* J. Am. Chem. Soc. *94*, 5125 (1972).
170a. *Stynes, D. V., Stynes, H. C., James, B. R., Ibers, J. A.:* J. Am. Chem. Soc. *95*, 1796 (1973).
170b. — — *Ibers, I. A., James, B. R.:* J. Am. Chem. Soc. *95*, 1142 (1973).
171. *Sutin, N., Christman, D. R.:* J. Am. Chem. Soc. *83*, 1773 (1961).
172. *Tang, S. C., Weinstein, G. N., Holm, R. H.:* J. Am. Chem. Soc. *95*, 613 (1973).
173. *Tenhunen, R., Marver, H., Pimstone, N. R., Trager, W. F., Cooper, D. J., Schmid, R.:* Biochemistry *11*, 1716 (1972).
174. *Tokel, N. E., Katović, V., Farmery, K., Anderson, L. B., Busch, D. H.:* J. Am. Chem. Soc. *92*, 400 (1970).
175. *Treibs, A.:* Liebigs Ann. Chem. *728*, 115 (1969).
176. — *Häberle, N.:* Liebigs Ann. Chem. *718*, 183 (1968).
177. *Tsai, R., Yu, C. A., Gunsalus, I. C., Peisach, J., Blumberg, W., Orme-Johnson, W. H., Beinert, H.:* Proc. Nat. Acad. Sci., USA, *66*, 1157 (1970).
178. *Tsutsui, M., Ostfeld, D., Francis, J. N., Hoffman, L. M.:* J. Coord. Chem. *1*, 115 (1971).
179. — — *Hoffman, L. M.:* J. Am. Chem. Soc. *93*, 1820 (1971).
180. — *Velapoldi, R. A., Hoffman, L., Suzuki, K., Ferrari, A :* J. Am. Chem. Soc. *91*, 3337 (1969).
181. — — *Suzuki, K., Koyano, T.:* Angew. Chem. *80*, 914 (1968).
182. — — — *Ferrari, A.:* J. Am. Chem. Soc. *90*, 2723 (1968).

183. — — — *Vohwinkel, F., Ichikawa, M., Koyano, T.:* J. Am. Chem. Soc. *91*, 6262 (1969).
184. *Tu, S. J., Tan, Y. J., Wang, J. H.:* Bioinorg. Chem. *1*, 79 (1971).
185. — *Wang, J. H.:* Biochem. Biophys. Res. Comm. *36*, 79 (1969).
186. *Ullrich, V.:* Angew. Chem. *84*, 689 (1972) Rev.
187. *Umrikhina, A. V., Yusupova, G. A., Krasnovkii, A. A.:* Dokl. Akad. Nauk. SSSR *175*, 1400 (1967).
188. *Wade, R. S., Castro, C. E.:* J. Am. Chem. Soc. *95*, 231 (1973).
189. — — J. Am. Chem. Soc. *95*, 226 (1973).
190. *Walker, F. A.:* J. Am. Chem. Soc. *92*, 4235 (1970).
190a. — J. Am. Chem. Soc. *95*, 1150, 1154 (1973).
191. *Wang, J. H.:* J. Am. Chem. Soc. *80*, 3108 (1958).
192. — Proc. Natl. Acad. Sci. *62*, 653 (1969).
193. — Accts. Chem. Res. *3*, 90 (1970).
194. — In: Oxygenases, p. 470 (*Hayaishi, O.*, ed.). New York: Academic Press 1962.
195. *Wasser, P. K.-W., Fuhrhop, J.-H.:* Ann. N. Y. Acad. Sci. *206*, 533 (1973).
196. *Waterman, M. R., Mason, I. S.:* Arch. Biochem. Biophys. *150*, 57 (1972).
197. *Wayland, B. B., Mohajer, D.:* Chem. Comm. *1972*, 776.
197a. — — Inorg. Nucl. Letters *9*, 633 (1973).
198. *Weaver, E.:* Ann. Rev. Plant. Physiol. *19*, 283 (1968).
198a. *Webb, L. E., Fleischer, E. B.:* J. Chem. Phys. *43*, 3100 (1965).
199. *Weiss, C., Kobayashi, H., Gouterman, M.:* J. Mol. Spec. *16*, 415 (1965).
200. *White, R. A., Tollin, G.:* J. Am. Chem. Soc. *89*, 1253 (1967).
201. *Whitlock, jr., H. W., Bower, B. K.:* Tetrahedron Letters *1965*, 4827.
202. *Whitten, D. G., Baker, E. W., Corwin, A. H.:* J. Org. Chem. *28*, 2363 (1963).
203. — *Yan, J. C., Carroll, F. A.:* J. Am. Chem. Soc. *93*, 2291 (1971).
204. *Wicholas, M., Mustacich, R., Jayne, D.:* J. Am. Chem. Soc. *94*, 5418 (1972).
205. *Williams, R. J. P., Smith, D. W.:* Structure Bonding *7*, 1 (1969).
206. *Wilson, L. D., Harding, B. W.:* Biochemistry *9*, 1615 (1970).
207. — *Dutton, P. L.:* Biochem. Biophys. Res. Comm. *39*, 59 (1970).
208. — — Arch. Biochem. Biophys. *136*, 583 (1970).
209. *Winfield, M. E.:* J. Mol. Biol. *12*, 600 (1965).
210. *Wittenberg, J. B., Wittenberg, B. A., Peisach, J., Blumberg, W. E :* Proc. Natl. Acad. Sci. USA, *67*, 1846 (1970).
211. *Wolberg, A., Manassen, J.:* Inorg. Chem. *9*, 2365 (1970).
212. — — J. Am. Chem. Soc. *92*, 2982 (1970).
213. *Woodward, R. B.:* Angew. Chem. *72*, 651 (1960).
214. — Ind. Chim. Belge *1962*, 1293.
215. *Yamamoto, S., Hayaishi, O.:* J. Biol. Chem. *242*, 5260 (1967).
216. *Yamamoto, A., Phillips, L. K., Calvin, M.:* Inorgan. Chem. *7*, 847 (1968).
217. *Yonetani, T., Drott, H. R., Leigh, J. S., Reed, G. H., Waterman, M. R., Asakura, T.:* J. Biol. Chem. *245*, 2998 (1970).
218. *Yoshida, Z., Ogoshi, H., Omura, T., Watanabe, E., Kurosaki, T.:* Tetrahedron Letters *1972*, 1077.
218a. *Zeller, M. V., Hayes, R. G.:* J. Am. Chem. Soc. *95*, 3855 (1973).
219. *Zerner, M., Gouterman, M.:* Theoret. Chim. Acta *4*, 44 (1966).
220. — — Inorg. Chem *5*, 1707 (1966).
221. — *Kobayashi, H.:* Theoret. Chim. Acta *6*, 363 (1966).

Received August 31, 1973

Optical Activity of Conjugated Proteins

G. Blauer

Department of Biological Chemistry, The Hebrew University, Jerusalem, Israel

Table of Contents

Abbreviations used:

ORD, optical rotatory dispersion
CD, circular dichroism
MORD, magnetic optical rotatory dispersion
MCD, magnetic circular dichroism
EPR, electron paramagnetic resonance
NMR, nuclear magnetic resonance
DBM, Debye-Bohr magneton
ADP, adenosine diphosphate
ARN, adenosine-5′-nicotinate
FAD, flavin-adenine dinucleotide
FMN, flavin mononucleotide
GMP, guanosine monophosphate
GTP, guanosine triphosphate
NAD, nicotinamide-adenine dinucleotide; (NAD^+, oxidized form; NADH, reduced form).
DNA, deoxyribonucleic acid
RNA, ribonucleic acid
BSA, bovine serum albumin
HSA, human serum albumin
HRP, horse-radish peroxidase
TMV, tobacco mosaic virus

I. Introduction

The elucidation of protein structure in solution is essential for a deeper understanding of protein function in biological systems. Methods and concepts developed during the past decades in macromolecular chemistry included those relevant to this important group of biopolymers, the proteins. While hydrodynamic methods yielded information about the overall shape of these macromolecules, certain optical methods, developed mainly during the past two decades, enabled investigators to obtain a wealth of data on the fine structure of proteins and other macromolecules in solution. A relevant and sensitive method used frequently is the measurement of optical activity. This method has been extensively applied to proteins and model polyamino acid systems from the late fifties, when new relationships between their structure and optical properties were recognized [for reviews on this subject, see for example, (*1*—*6*)]. The advance made in theory (*7*) [see reviews (*6, 8, 9*)], enabled deductions to be made on the occurrence of ordered and random conformations and other structural features in solution. Recent developments in instrumentation (see below: Methods of Measurement) permitted measurements of both optical rotatory dispersion (ORD) and circular dichroism (CD) to be made in dilute solutions over a much wider range of wavelengths and under conditions of larger absorbance values. Indeed, a very steep increase in the amount of work on various aspects of the optical activity of biopolymer systems is evident from a survey of the literature of the past 10—15 years.

This review is concerned with the optical activity of systems containing proteins combined with non-protein compounds, frequently referred to as "conjugated proteins". These systems include, *inter alia*, naturally occurring enzyme-coenzyme complexes or hemoproteins. The present survey also includes in this category artificial complexes between proteins and dyes or other non-protein compounds. A separate treatment of the optical activity of "conjugated proteins" seems justified at present because of the many common features exhibited by various such systems of biological and chemical importance. Unconjugated proteins do not have chromophores with electronic transitions above 300 nm, whereas most of the conjugated molecules investigated absorb in the visible or near-ultraviolet regions where transitions may become optically active. In many cases, either small differences or none at all can be observed between a conjugated protein and its corresponding apoprotein in the protein bands below 300 nm. It will be shown that for many conjugated proteins, some specific deductions about stereochemical and environmental features are possible. In some cases, these may be correlated with functional properties of the systems studied.

Obviously, the structural information gained by ORD or CD measurements and the corresponding electronic absorption spectra is specific but limited. Whenever possible, they should be combined with complementary data such as those obtained by X-ray diffraction, fluorescence measurements, NMR, EPR and other magnetometric methods, hydrodynamic measurements, relaxation methods as well as immunochemical methods.

Several recent reviews on optical activity of biological systems also include some conjugated proteins and properties of coenzymes or model compounds [see, for example, (*10–14*)]. In the present survey, particular emphasis is placed on systems investigated recently, many of which are not included in other pertinent reviews. On the other hand, this review may not give a completely comprehensive account of all the work published on the optical activity of conjugated protein systems. The main purpose is rather an evaluation of recent information obtainable on the structure of both natural and artificial protein complexes and compounds as measured by ORD or CD.

Apart from protein systems, the optical activity of various other substances and systems of biological interest has also been investigated recently, such as nucleic acids, polysaccharides, ribosomes or membrane fragments [see, for example, (*11–13, 15–17*)]. Although these are not included in the present review, many analogies with regard to the generation of optical activity and the interpretation of data can be found between these and protein systems.

II. Some Fundamental Concepts of Optical Activity[1])

Linearly polarized light has its electric vector, $\vec{E}$, oscillating in a given plane. This plane-polarized wave may be formed by superposition of right-and left-circularly polarized waves. In a given medium, the velocities of propagation of these two components, $\vec{E}_L$ and $\vec{E}_R$, will be different if the refractive index for one of the components differs. This situation will occur in the case of dissymmetric molecules interacting with the plane-polarized light. As a consequence of this difference in velocities, the resultant vector $\vec{E}$ will now oscillate in a new plane, rotated by an angle α', relative to the original plane of oscillation. At a given (vacuum) wavelength λ, the angle α'_λ per unit length of optical path is proportional to the difference in the refractive indices for left-and right-circularly polarized light:

$$\alpha'_\lambda = \frac{\pi}{\lambda}(n_L - n_R) \qquad (1)$$

1) See reviews (*18–25, 6, 8, 12, 17*).

For an asymmetric substance in solution, this measure of optical activity is conveniently expressed in terms of the specific rotation $[\alpha]_\lambda$:

$$[\alpha]_\lambda = \frac{\alpha'_\lambda}{c'} \frac{1800}{\pi} \tag{2}$$

where c' is the concentration of the optically active solute in g per cm^3 and the numerical factor converts radians per cm into the experimental unit of degrees per dm.

For comparison of different substances, it is convenient to use the molar rotation, $[\phi]$. This is defined as follows

$$[\phi]_\lambda = [\alpha]_\lambda \frac{M}{100} = \frac{10\alpha_\lambda}{Cl} \tag{3}$$

where M is the molecular weight of the substance under consideration; α_λ, the observed rotation in degrees; C, the concentration in moles per liter; l, the optical path length in dm.

Because of absorption, the intensity of the plane-polarized light will also decrease exponentially with distance. In optically active media, the two components $\vec{E}_L$ and $\vec{E}_R$ will, therefore, not only show different velocities but they will also be absorbed to a different extent. This phenomenon is termed circular dichroism (CD). The head of the resulting vector $\vec{E}$ will appear to move on an ellipse. Thus, elliptically polarized light is produced from linearly polarized light in a medium showing circular dichroism.

By analogy with molar rotation, the molar ellipticity $[\theta]_\lambda$ is defined by

$$[\theta]_\lambda = \frac{\theta'_\lambda}{c'} \frac{1800}{\pi} \frac{M}{100} = \frac{10\,\theta_\lambda}{Cl} \tag{4}$$

where θ' is the arctg of the ratio of the minor to the major axis of the ellipse, which is usually very elongate; θ is the observed ellipticity in degrees.

The conventional units of both $[\phi]$ and $[\theta]$ are degrees $\cdot$ cm^2 $\cdot$ decimole^{-1}. It follows from the absorption laws that

$$[\theta]_\lambda = 3300\,(\varepsilon_\lambda^L - \varepsilon_\lambda^R) \tag{5}$$

where ε^L and ε^R are the molar extinction coefficients for right- and left-handed circularly polarized light, respectively. Under certain conditions, the values of $[\phi]_\lambda$ or $[\theta]_\lambda$, measured in given media, may be reduced to *vacuum* by multiplication by the Lorentz factor $\frac{3}{n_\lambda^2 + 2}$, where n_λ is the refractive index of the medium.

The dependence of the optical rotation on wavelength (or wavenumber) is termed optical rotatory dispersion. Both ORD and CD spectra show their largest changes near an electronic absorption band. These "Cotton effects" are shown schematically for a single transition in Fig. 1.

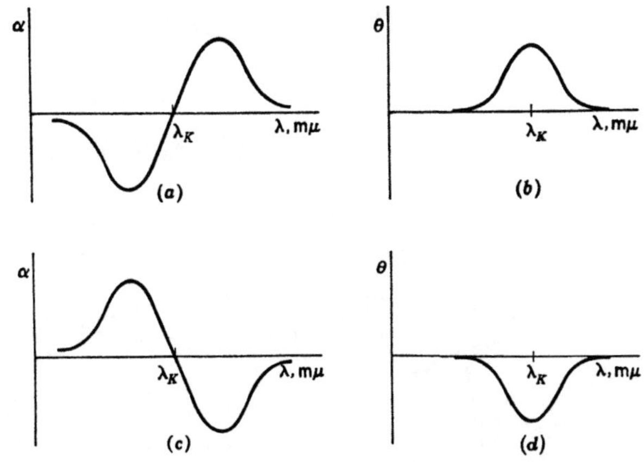

Fig. 1. Behavior of α and θ in the immediate vicinity of an isolated absorption band (idealized): (a) and (b) positive Cotton effect, (c) and (d) negative Cotton effect. [Reproduced from Ref. (*18*)]

The Cotton effects can be either positive or negative. ORD and CD are related phenomena and data obtained by each method can be transformed to the other by means of the Kronig-Kramers relations [*e.g.*, Ref. (*18*)]

$$[\phi_k(\lambda)] = \frac{2}{\pi} \int_0^\infty [\theta_k(\lambda')] \frac{\lambda'}{\lambda^2 - \lambda'^2} d\lambda' \tag{6}$$

$$[\theta_k(\lambda)] = \frac{2}{\pi\lambda} \int_0^\infty [\phi_k(\lambda')] \frac{\lambda'^2}{\lambda^2 - \lambda'^2} d\lambda' \tag{7}$$

Cotton effects of systems involving several optically active transitions are superpositions of partial rotations or ellipticities. These may be resolved from the experimental data into their components (*k*) (see Fig. 2). Computer programs can be used for this purpose. Both light-

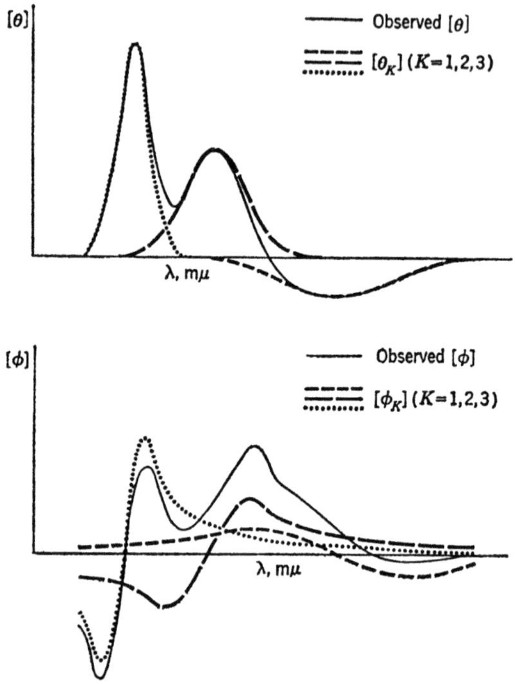

Fig. 2. (Top). [θ] as the superposition of three partial molar ellipticities. (Bottom). [ϕ] as the superposition of three partial molar rotations. [Reproduced from Ref. (*18*)]

absorption and CD bands are usually approximated by Gaussian curves. It is often difficult to achieve a unique resolution [*e.g.*, Refs. (*26, 12*)].

Parameters characterizing either single light-absorption or CD bands are the positions of the band extrema λ_k^0; the corresponding extinction or ellipticity values, ε_k^0 or $[\theta_k^0]$, respectively; and the band half-widths Δ_k given at ε_k^0/e or $[\theta_k^0]/e$, respectively. The integrated band areas are important quantities. For ordinary light absorption they are a measure of the intensity of the transition given either as oscillator strength, or dipole strength D_k

$$D_k \simeq \frac{(2303)\, 3\, hc}{8\, \pi^3\, N} \int_0^\infty \frac{\varepsilon_k(\lambda)\, d\lambda}{\lambda} \tag{8}$$

where h is Planck's constant; c, velocity of light; N, Avogadro's number. The following approximation is frequently used for a Gaussian absorption band (*18*)

$$\mathrm{D}_k \simeq 1.63 \times 10^{-38} \frac{\varepsilon_k^0 \Delta_k}{\lambda_k^0} \quad \text{(cgs)} \tag{9}$$

The dipole strength is the value of the square of the electric dipole moment $\vec{\mu}_k$ of a given transition

$$\mathrm{D}_k = \mu_k^2 \tag{10}$$

The transition moment is a vector having a fixed orientation within a chromophore. It depends on the wave functions of the ground and excited states.

By analogy with the dipole strength [Eq. (8)], the rotatory (or rotational) strength R_k of a transition can be obtained from the integrated ellipticity values of a dichroic band. R_k may be evaluated approximately for a Gaussian curve by [see Ref. (*18*)]

$$\mathrm{R}_k \simeq 1.23 \times 10^{-42} \frac{[\theta_k^0]\Delta_k}{\lambda_k^0} \quad \text{(cgs)} \tag{11}$$

This fundamental quantity, R, in optical activity can be expressed as the imaginary part of the scalar product of electric and magnetic dipole moments of a transition

$$\mathrm{R}_k = \mathrm{Im}\,(\psi_0|\vec{\mu}|\psi_k) \cdot (\psi_k|\vec{\mathrm{m}}|\psi_0) \tag{12}$$

R vanishes for a symmetrical molecule. In this case, either one of the transition dipole moments is zero or they are perpendicular to each other.

A natural and frequently used unit for R is the Debye-(Bohr) magneton ($\simeq 0.927 \times 10^{-38}$ cgs units).

The anisotropy factor of a transition, g_k, is defined by

$$g_k = 4\frac{\mathrm{R}_k}{\mathrm{D}_k} \tag{13}$$

It may assist in the assignment of a transition.

For interpretation of the generation of optical activity, three complementary types of asymmetric interactions may be considered [see Ref. (*8*)], using as a simple model a molecule having two groups and two electronic transitions:

a) Two proximate groups, each of which has an electronic transition dipole moment, are coupled [coupled oscillator model (*27*)]. For identical groups, the excited state is doubly degenerated [*Moffitt* (*7*)]. This can be treated in terms of exciton splitting observed in molecular crystals [*e.g.*

(*28*), and refs. cited therein]. In this case, the wave function of the excited state is

$$\psi'_{\pm} = \frac{1}{\sqrt{2}} (\phi_1 \phi_2' \pm \phi_1' \phi_2) \tag{14}$$

where the excited and ground states are indicated by primed and unprimed functions, respectively. The rotatory strength of the two separated states is given by

$$R = \pm \frac{\pi}{2\lambda} (\vec{r}_{12} \cdot \vec{\mu}_1 \times \vec{\mu}_2) \tag{15}$$

where $\vec{r}_{12}$ is the vector distance between the electric transition moments $\vec{\mu}_1$ and $\vec{\mu}_2$. λ refers to the transition of a non-interacting group.

In this case the dichroic bands (and ORD curves) will show the general shape indicated in Fig. 3.

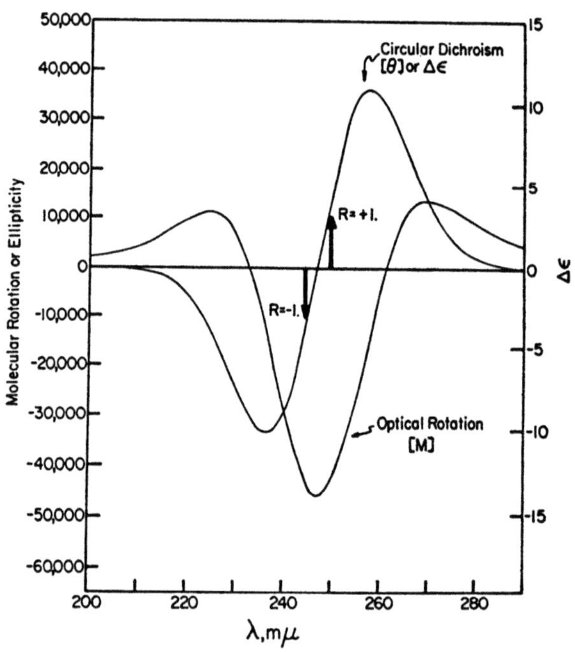

Fig. 3. A rotatory strength "couplet" is a characteristic form of dispersion which arises when two Cotton effects of equal but opposite rotatory strength lie less than a band width from one another. In the figure, $R_1 = 1$ DBM, $\Delta_1 = 15$ nm, $\lambda_1 = 250$ nm; $R_2 = -1$ DBM, $\Delta_2 = 15$ nm, $\lambda_2 = 245$ nm. A couplet is an extremely effective way of detecting small band splittings. [Reproduced from Ref. (*8*)]

Alternatively, the sign of both exciton bands may be reversed. For the more general, nondegenerate case of coupled electric dipoles involving two different chromophoric groups, the rotatory strength R is given by

$$R = \pm \frac{2\pi}{hc} \frac{V_{12}E_1E_2}{(E_2^2 - E_1^2)} (\vec{r}_{12} \cdot \vec{\mu}_1 \times \vec{\mu}_2) \tag{16}$$

where V is the perturbation energy and E is the energy of a state.

It can be shown that

$$\sum_i R_i = 0 \tag{17}$$

This is in accordance with the known sum rule [*Kuhn* (*29*)] stating that the sum over all rotatory strengths in a molecular system gives zero.

b) According to the one-electron theory [*Condon et al.* (*30*)], a single chromophore having one electric and one magnetic transition is considered. These are mixed by the perturbing field of the environment with some loss of symmetry in the chromophore.

c) An electric transition in one group interacts with a magnetic transition of another (*8*).

In both cases (b) and (c) the rotatory strength will be expressed in terms of the energies and of the scalar products of the electric and magnetic transition moments involved.

A useful classification of chromophores with regard to optical activity distinguishes between two main groups (*18*):

a) Inherently symmetric chromophores which are asymmetrically perturbed by their environment. A classical example is given by the carbonyl chromophore in steroids.

b) Inherently dissymmetric chromophores which may generate very large rotatory strengths. A known example is resolved hexahelicene, which consists of six fused benzene rings in a helical type of arrangement, producing a large and dissymmetric chromophore.

Electric dipole coupling between large and dissymmetrically oriented transition moments can also lead to generation of very large rotatory strengths. Some examples will be given below.

Symmetry rules for determination of the signs and orders of magnitude of Cotton effects have been deduced [see Ref. (*8*)].

In protein systems in particular, a distinction between "intrinsic" and "extrinsic" Cotton effects has been made (*31*). The "intrinsic" effects are the result of internal dissymmetric interactions of protein chromophores such as the amide transitions in helical arrays. "Extrinsic" effects relate to non-protein substances which are usually optically inactive but exhibit optical activity when conjugated to a protein or to an asymmetric polymer. For the commonly observed case of mutual electric dipole coupling between transition moments of the conjugated substance and those of the macromolecule, the term "extrinsic Cotton effect" appears to be defined clearly. However, the optical activity observed may, in some cases, be due to interaction between the conjugated molecules or molecular groups bound asymmetrically on the matrix of the asymmetric macromolecule. The corresponding Cotton effects may then be regarded as "intrinsic" with respect to the bound molecules. It is suggested that, whenever possible, the actual mechanism leading to optical activity be used for description of the system rather than a general and ambiguous terminology.

III. Methods of Measurement

In recent years great advance has been made in instrumentation for ORD and, in particular, for CD measurements [see, for example, Refs. (*32–39*, *17*)]. The improvements in measurement of CD within an absorption band enabled progress in the determination of the optical activity of conjugated protein molecules, particularly for systems of relatively small activity. A frequently used calibration substance for CD instruments is (+)-camphor-10-sulfonic acid in aqueous solution. In order to achieve correct calibration, optical rotation values of the calibration substance should be transformed to CD values used for standardization (*40*, *41*).

The choice between ORD and CD as a method of measurement will depend on the information desired in each case. Dichroic bands usually fall off with wavelength as curves of Gaussian type, while corresponding optical rotation curves decay less steeply at a distance from the band center because of the dependence $1/(\lambda^2 - \lambda'^2)$ (see Fig. 1). For conjugated protein systems, CD measurements will be of considerable advantage over ORD throughout most of the spectral range. In particular, above approximately 300 nm, the unconjugated proteins will show only negligible ellipticity values under the usual conditions of measurement, while many of the biologically important conjugated molecules have optically active transitions in the visible and near-ultraviolet regions. CD data measured in these regions will, therefore, be due to the optical activity of the conjugated molecules and can be expressed on a molar basis of the latter. On the other hand, ORD data of the same systems would include background rotation of the protein component which, in some cases, may overshadow rotation effects due to the conjugated molecules. In the far ultraviolet region, however, the main contribution to the optical activity of a conjugated protein will be made, in many cases, by the peptide transitions of the protein only. In this case, the optical activity can be expressed as a mean residue rotation or ellipticity of the protein, provided transitions of the conjugated molecules do not directly affect these values. ORD data will be valuable in cases where measurements at a distance from the center of an absorption band are desirable, as for example, for evaluation of the parameter b_0 in the Moffitt equation (see below). Other transitions except those of the peptide groups should, of course, not interfere.

In the absorption region of aromatic amino acid residues (about 250—300 nm), CD may again be a preferred method, particularly if only small changes in optical activity occur upon conjugation to a protein. In many cases, the ratio of ellipticity to absorption is relatively small in this spectral region and both measurement and interpretation of effects require particular care.

Magnetic ORD and CD. Applications of the anomalous dispersion of the Faraday effect for the study of proteins are still limited. For general experimental and theoretical aspects of MORD and MCD and further references see Refs. (*42—49*). [See below several investigations by these methods on myoglobin, hemoglobin or cytochrome *c*].

IV. Optical Activity of Proteins

A brief survey on the optical activity of unconjugated proteins in solution is essential for the topic of the present review, although extensive reviews are available on the subject of ORD and CD of proteins and polypeptides [*e.g.*, (*1—3, 37, 50—52, 15, 6, 11, 17*)].

A considerable amount of work on correlations between the optical activity of the far ultraviolet peptide amide bands (n—π^* near 222 nm and π—π^* near 206 and 190 nm) and ordered structures in proteins and polypeptides has accumulated over the past 10—15 years. A significant advance in the evaluation of secondary protein structure has been made by application of the Moffitt equation [Ref. (7)] to the ORD of protein systems in solution:

$$[\phi']_\lambda = \frac{a_0\,\lambda_0^2}{\lambda^2-\lambda_0^2} + \frac{b_0\,\lambda_0^4}{(\lambda^2-\lambda_0^2)^2} \tag{18}$$

where $[\phi']_\lambda = [\alpha]_\lambda \frac{M_0}{100} \frac{3}{n_\lambda^2+2}$ is the effective reduced residue rotation; M_0 is the molecular weight of a residue; a_0, b_0, and λ_0 are constants.

This equation was originally based on exciton splitting of amide π—π^* bands in a helical array [see a discussion, Ref. (*2*)]. It is applied in a wavelength region remote from the absorption band center. The parameter b_0 has empirically been used as a measure of α-helix content. It is usually taken to be —630 (deg · cm^2 · decimole^{-1}) for a complete right-handed α-helix (*53*) and zero for a random conformation, assuming b_0 to be a linear function of helix content. A modification of this relationship has been suggested by *Shechter* and *Blout* (*54*) [see also (*55, 5*); see, however, (*56, 51*)]. A number of difficulties experienced in many cases render this method, based on the Moffitt equation, only a semiquantitative tool or even make conformational deduction impossible (*11*). For example, there may be contributions from other structures, dependence on the length of the helix for short helices, (*57, 58*), aggregation effects [see Refs. (*59, 51*) and refs. cited therein] or contributions from protein transitions centered in the region 250—300 nm. In conjugated protein or polypeptide systems, overlapping rotations which originate from transitions of the conjugated molecules may affect calculations of helix content.

The ORD spectra near 233 nm (trough) and the CD bands near 222 nm (band minimum) have also been used extensively for estimating ordered protein or polypeptide structures.

Frequently used relationships for calculation of α-helix content are [see (*51, 59, 15, 60, 61, 62*)]:

$$\text{Fraction helix} \simeq -([\phi']_{233} + 2000)/13000 \qquad \text{(ORD)} \qquad (19)$$

$$\text{Fraction helix} \simeq -([\theta]^{\mathrm{M_0}}_{222-223} - 4000)/40000 \qquad \text{(CD)} \qquad (20)$$

The parameters given may differ to some extent for various systems, since these values were obtained for synthetic polyamino acids and may not be applicable to many protein systems [*e.g.*, (*63, 59, 64, 60, 65—68*)].

Moreover, all the complicating factors mentioned above in connection with the use of b_0 for determination of helix content may also affect the estimates by Eqs. (19) and (20). This includes overlapping effects of transitions of molecules conjugated to proteins; however, at a distance from the active band center their influence would, in general, be smaller in CD spectra because of their steeper wavelength dependence as compared to ORD. Among possible contributions to the optical activity by other structures, β-structures may be prominent in proteins [*e.g.*, (*15, 61, 69*) and refs. cited therein]. Fig. 4 illustrates far ultraviolet CD spectra of polyamino acids in different conformations.

Recently, simultaneous evaluation by CD (190—250 nm) of α-helical, random and β-structures has been carried out with the aid of computer programs, using appropriate parameters and results from X-ray investiga-

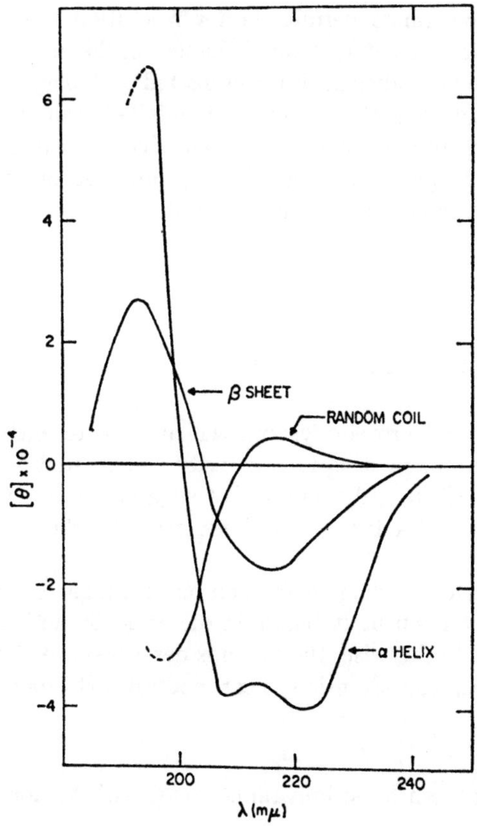

Fig. 4. Circular dichroism spectra of poly-α, L-glutamic acid in α-helical and random-coil forms and of poly-L-lysine in β-structure (β-sheet). [Reproduced from Ref. (*67*)]

tions on various proteins [parameters based on poly-L-lysine (*70*); on poly-L-serine-8 M LiCl (*71*); on infrared data (*72*); computations by parameters based on X-ray analysis of proteins (*69, 68*)]. More than three sets of parameters may be necessary for curve fitting (*73*).

The above short survey on the far ultraviolet optical activity of proteins demonstrates the difficulties encountered in quantitative correlations between observed data and protein structures in solution; nevertheless, conclusions may be reached in many cases, although with great caution and reservations.

In the spectral range of 240—300 nm, contributions to the optical activity of a protein may originate from transitions of side chains of aromatic amino acids or of disulfides [see also Ref. (*74*)]. Measurement of

CD in the above range permits better resolution than of ORD. CD-difference spectra may detect small changes in the protein bands, occurring under various conditions. For conjugated protein systems, differences in the optical activity of the protein bands between the unconjugated and conjugated protein may be a sensitive indicator for structural changes in the biopolymer upon binding, provided direct contributions from the bound molecule can be ruled out.

V. Conjugated Protein Systems

Various conjugated protein systems which have recently been investigated by measurement of optical activity are treated briefly below. In most cases described, only qualitative suggestions about the origin of the observed optical activity have been made by the respective investigators.

For proteins containing more than one conjugated ion or molecule, their classification is usually based on the optically active moiety.

Unless stated otherwise, the systems described are close to the usual conditions of aqueous solutions near neutral *p*H, room temperature, etc.

A. Complexes of Synthetic Polyamino Acids with Various Molecules

These systems will be dealt with first since they may be regarded as models for conjugated proteins. The optical activity of complexes between poly-α,L-glutamic acid and the symmetrical, cationic dye acridine orange has been extensively studied (*75, 76*); [see also Refs. (*77, 78*) and refs. cited therein for complexes involving other acridine derivatives and dyes]. These complexes show large Cotton effects in the dye absorption region when the poly-L-glutamic acid is in α-helical conformation (below *p*H 5). The sign of these Cotton effects is reversed for poly-D-glutamic acid, which is considered to have a helical conformation of opposite screw sense. Complexes involving acridine orange and non-helical polyglutamic acid (higher *p*H values) did not show similar Cotton effects. However, it has been subsequently shown (*79*) that the mode of preparation of the complex may determine the optical activity observed and that, under certain conditions, Cotton effects can also be produced with non-helical polyglutamic acid [see also Refs. (*80, 81*)]. This result may be of general relevance to conjugated biopolymer systems as it may indicate the possible existence of small regions of (dissymmetric) order in a macromolecule assumed to be in a "random" conformation. The Cotton effects observed below *p*H 5 have been considered to result from helical dye aggregation on the polypeptide helix, allowing interactions between the transition moments of the dye molecules (*76*). Formation of a left-handed superhelix of aggregated acridine orange molecules on the right-handed α-helix of poly-α,L-glutamic acid has been inferred from spectroscopic data (*81*). It has been reported that a single symmetrical dye molecule bound covalently to dissymmetric polybenzyl-L-glutamate also shows Cotton effects in the region of its absorption bands (*31*).

Helical poly-α,L-lysine at *p*H 11 forms a "red complex" (hemochrome type) with ferriprotoporphyrin IX, the prosthetic group of hemoproteins, while with non-helical polylysine a "green complex" is formed below about *p*H 10 (*82, 83*). A large negative Cotton effect is observed in the Soret region for the "red complex", but this effect has been reported to be absent at lower *p*H (*84*). Free ferriheme is optically inactive, but ferriheme bound to helical poly-L-lysine is considered to be arrayed between the polypeptide helices (*85, 83*). In this arrangement, dipole coupling between neighboring and dissymmetrically oriented heme transition moments may account for at least part of the Cotton effects observed [see also Ref. (*37*) for the complex ferriheme-poly-L-histidine]. Other possible contributions originating from other heme interactions may also play a role. Similar Cotton effects have been measured in the system ferriheme-poly-α,L-ornithine at *p*H 11.3—11.4 (*86*). Another interesting observation made on this system is that the rotation trough changes considerably at 233 nm to more negative values upon addition of ferriheme to the polyamino acid. This change suggests an increase in the apparent α-helix content from about 16% to 32%, below a molar ratio of ornithyl residues: ferriheme of about 40 (*86*). Some analogies between these model systems and hemoproteins have been pointed out (see below, myoglobin or hemoglobin). Under comparable conditions, poly-α,L-lysine has a much higher α-helix content than poly-α,L-ornithine (*87*), and no significant changes of the rotation at 233 nm have been observed upon addition of ferriheme to helical poly-L-lysine (*86*).

Interactions between basic polyamino acids such as poly-L-lysine and various nucleotides, such as GMP, were followed by ORD (*88*) and CD (*89*) in the ultraviolet region, estimating conformational changes in the complex components.

Binding between DNA and poly-L-lysine at *p*H 7 was monitored at various molar ratios of the complex components by measurement of ORD changes in the range 220 to 360 nm. These changes could be compared with those occurring in DNA-protein complexes (see below, nucleohistones). Possible interpretations included tilting of the bases without perturbation of the interstrand hydrogen bonding in DNA, or changes in the tertiary structure (super-coiling) (*90*). Analogous and even larger effects shown by the above system were considered to be due to possible formation of a crystalline phase (*91*, *92*). The effects of acetylation of poly-L-lysine on the CD spectra of its complexes with DNA were studied (*93*).

B. Metalloproteins[2)]

1. Hemocuprein

A combination of spectroscopic methods (EPR, light absorption and CD) was used in order to obtain information on the copper binding site of bovine hemocuprein (erythrocuprein), which has been considered to be intermediate in its spectroscopic properties between the "blue" copper proteins and complexes of copper with model peptides (*94*).

2. Bovine Erythrocyte Copper-zinc Protein

This metalloprotein exhibited several positive CD bands in the range 300—600 nm which disappeared upon removal of the metal atoms. Simultaneous changes were observed in the "aromatic region" and also small changes in the far-ultraviolet region, in which very little or no α-helical structures were indicated (*95*).

3. Ceruloplasmin[3)]

Native porcine ceruloplasmin showed small CD bands in the visible region, which disappeared on removal or reduction of the copper. Far ultraviolet CD and ORD data indicated the presence of β and random structures (*97*). Several CD bands were recorded for human ceruloplasmin (*p*H 5.5) in the range 300 to 1100 nm. Spectral changes were observed upon addition of azide (*98*).

4. Hemocyanins

Light-absorption and CD studies of hemocyanins from two different molluscan species showed several bands in the visible region characteristic also of other copper proteins. The greater resolving power of CD spectra yielded more accurate information than previous ORD measurements. *d-d* Transitions with distorted coordination sites could account for the absorption intensities observed without invoking charge transfer. By comparison of the CD data with those of peptide-Cu(II) complexes, the

---

2) For reviews on earlier data, see also Refs. (*10*, *14*).
3) See also, Ref. (*96*).

possibility of histidyl residues acting as ligands for the copper was suggested [Ref. (*99*) and refs. cited therein]. The characteristic CD bands in the visible region of *Octopus vulgaris* hemocyanin and a large band at 350 nm (see Fig. 5) disappeared upon removal of copper or oxygen

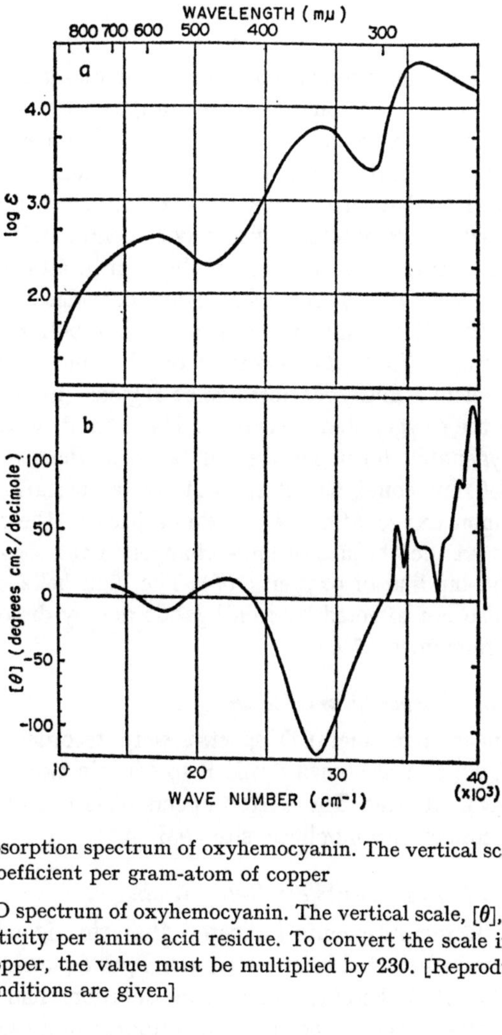

Fig. 5a. The absorption spectrum of oxyhemocyanin. The vertical scale is expressed as extinction coefficient per gram-atom of copper

Fig. 5b. The CD spectrum of oxyhemocyanin. The vertical scale, $[\theta]$, is expressed as the molar ellipticity per amino acid residue. To convert the scale into gram-atom ellipticity of copper, the value must be multiplied by 230. [Reproduced from Ref. (*100*) where conditions are given]

(100). A comparison of the CD spectra in the range 260—700 nm of molluscan and arthropod hemocyanins showed large differences which were attributed to fundamental differences in the oxygen binding sites of these related proteins (*101*). The deoxy- and oxy-forms of the hemocyanin from the squid showed some differences in both sedimentation coefficients and CD spectra (250–280 nm). These were attributed to conformational changes occurring upon oxygenation since similar CD bands appeared in this region in the apoprotein (*102*). Analogous observations and conclusions resulted in the case of *Busycon* hemocyanin (*103*).

5. Hemerythrin

Preliminary ORD measurements on oxy- and methemerythrin (*Phascolosoma gouldi*) had demonstrated the appearance of visible-region Cotton effects upon oxygenation. These effects were considered to indicate changes in the relative orientation of the iron-ligand groups of the protein (*104*). CD and absorption spectra of deoxy-, oxy- and methemerythrin were used in conjunction with model compounds and magnetic measurements to interpret the state of the iron in all cases. Conformational changes around the iron were again considered to occur upon oxygenation (*105*). ORD and CD measurements of both oxy- and deoxyhemerythrin from *Sipunculus nudus* showed three negative dichroic bands in the visible and near ultraviolet regions (iron chromophores) in the case of the oxygenated derivative. These bands were again absent for the deoxygenated form. Differences between the two forms in the range 250–350 nm could be interpreted by reorientation of aromatic side chains upon oxygenation, while similarities in the far ultraviolet spectra indicated the absence of large changes in the secondary protein structure upon binding of oxygen (*106*). The high helix content of the protein was also not affected by iron ligands nor by dissociation of the proteins into monomers (*107*).

6. Plastocyanin (Chenopodium Album)

Visible and near-ultraviolet CD spectra were recorded for this blue plant protein, which was shown by EPR to contain two bivalent copper atoms in equivalent sites. The Cotton effects observed indicated asymmetry of the copper chromophoric site (*108*).

7. Stellacyanin (Rhus Vernicifera Blue Protein)

ORD and EPR measurements indicated that the ligand field of the single cupric ion is distorted from the regular square-planar configuration [(*109*) and refs. cited therein; similar considerations had been applied to the Cu(II) sites in the copper protein ceruloplasmin (*96*)]. This could also explain the very intense blue color observed. ORD data in the ultraviolet region indicated little or no helical structure in this plant protein (*109*). CD spectra of stellacyanin in the range 300 to 1100 nm were reported (*98*).

8. Blue Protein of Rice Bran

An ORD Cotton effect was observed near 450 nm (*110*).

9. Umecyanin

Light-absorption and CD data of this blue copper(II)-protein from horse-radish roots were measured and resolved in the range 200 to over 700 nm at various *p*H values. The relatively largest rotatory strength of the several bands observed appeared to be above 740 nm, which was the limit of the measurements. The spectra were similar to those of the *Pseudomonas* blue protein. The origin of the bands was not identified. Far ultraviolet data indicated very little α-helix content, about 40% β-structure and the remainder to be in random-coil conformations (*111*).

10. Pseudomonas Blue Protein and Polyporus Laccase

ORD and CD studies (190 to 700 nm) were carried out in order to obtain information about the environment of the chromophores. Above 300 nm, six overlapping Gaussian bands were resolved from both light-absorption and CD spectra. The data were checked by comparison of ORD measurements and calculations by Kronig-Kramers transforms. The exact nature of the multiple optically active transitions remains uncertain but the chirality of the copper environment is different for both proteins since the band signs differ (*112, 98*). ORD measurements on *Pseudomonas* azurins have also shown Cotton effects in the visible region, again indicating asymmetry at the copper-binding site (*113*).

11. Transferrin

CD studies of metal complexes (iron, copper) with human serum transferrin showed bands in the visible region (*114*) [earlier work in this field, mainly by ORD measurements is cited in Ref. (*114*)]. The binding of cupric or zinc ions to apotransferrin had no significant effect on the far ultraviolet CD (ORD) spectra (*114*). Similar results were also reported for chicken ovotransferrin (*115*). However, there were differences in the aromatic and other spectral regions up to 1000 nm between the CD spectra of the two copper transferrins indicating dissimilarities in the respective metal-binding sites (*115*).

12. Ferredoxin, Rubredoxin, Adrenodoxin

ORD spectra (300—600 nm) of bacterial ferredoxin and rubredoxin suggested similar asymmetric environments for both non-heme iron proteins. Spectral differences with plant ferredoxin were attributed to transitions involving the iron-sulfur chromophores (*116*). Subsequent CD measurements in the range 215—800 nm resolved more differences between oxidized bacterial ferredoxin and rubredoxin; however, the spectra were considered to be sufficiently alike to support the existence of similar optically active chromophores in both cases (*117*). Changes in the visible-range ORD spectra observed upon reduction of adrenodoxin from pig adrenal cortex have been correlated with structural changes around the iron occurring during the oxidation-reduction cycle (*118*). Spectroscopic data, which included a large anisotropy factor, indicate that the 1.6 μ CD band of rubredoxin originates from a magnetic dipole allowed transition in a tetrahedral ligand field of ferrous ion (*118a*). CD spectra of alfalfa ferredoxin were reported in the range 250—600 nm (*107*). Ferredoxin from spinach and adrenodoxin from beef adrenal mitochondria, which differ in their EPR signals, show similar CD spectra (300—700 nm) in both the oxidized and reduced forms, respectively. This similarity indicates that the environment of the iron chromophores is essentially identical in both proteins (*119*). The oxidized forms of spinach ferredoxin and of xanthine oxidase showed similarities in their CD spectra in the range 350—800 nm, suggesting that transitions involving iron are involved in both cases (*120*). A comparison of ORD and MORD (15 °C) of both oxidized and reduced spinach ferredoxin showed no drastic changes with regard to the ORD spectrum except for the appearance of additional rotations in MORD, which differed considerably for the oxidized and reduced forms (*121*). Spinach ferredoxin was also employed to demonstrate the use of MCD of a "marker" molecule such as a tryptophan residue in the presence of other chromophores. A marker group should have practically degenerate transitions and, in contrast to CD, the MCD will be relatively insensitive to the environment (*122*).

13. Xanthine Oxidase

Similarities between the CD spectra of both oxidized and reduced xanthine oxidase and corresponding ferredoxin spectra (350—700 nm) were shown (*123*). Xanthine oxidase from milk which contains FAD, molybdenum and iron was further investigated by CD and MCD (200—700 nm), and weak MCD bands at 520 and 375 nm were attributed to iron-sulfur and molybdenum chromophores, respectively (*48*).

14. Hydrogenases

The non-heme iron protein hydrogenase from *Clostridium pasteurianum* showed several CD bands in the visible and near ultraviolet regions in both oxidized and reduced forms. Similarities with the spectra of spinach ferredoxin and adrenodoxin suggest a similar structure of the metal-ligand chromophore in these proteins (*124*).

15. Ferritin

The iron-storage protein ferritin from horse spleen and its apoprotein were compared both in the absence and presence of aqueous guanidinium chloride, using far-ultraviolet CD, sedimentation velocity, electrophoresis and electron microscopy as methods. At neutral *p*H, apoferritin appeared to be more stable than the iron-containing protein, which dissociated into subunits in 7.0 M guanidinium chloride. As judged by CD measurements, extensive disruption of ordered structures occurred upon subunit formation, which could be reversed largely by removal of the denaturing agent (*125*).

16. *Metallothionein*

Both human and equine metallothionein exhibited a positive ORD Cotton effect at 254 nm, which has been considered to be due to asymmetric binding of cadmium to multiple sulfur ligands (*126*).

17. *Carboxypeptidase*

Replacement of the zinc by cobalt in either bovine procarboxypeptidase or carboxypeptidase resulted in a visible-range CD spectrum, which was inverted and increased in magnitude upon addition in glycyl-L-tyrosine. Correlations between the optical properties and the catalytic potential of the metalloproteins have been proposed (*127*).

18. *Carbonic Anhydrase*

In the 1:1 enzyme-inhibitor complex carbonic anhydrase-azosulfonamide, large Cotton effects have been induced in the visible region upon binding of the optically inactive dye (*128*) [for a similar system, see Ref. (*129*)]. Differences in CD spectra observed among various isozymes and species variants of carbonic anhydrase suggest conformational or structural differences at the respective active centers (*128*). 1:1 binding of the metal-chelating agent 4-(8-hydroxy-5-quinolylazo)-1-naphthalene sulfonate to bovine carbonic anhydrase B inhibits the enzyme and produces Cotton effects in the visible region, provided that one zinc ion is present. The chelating agent is considered to be bound in the vicinity of the active site (*130*). MCD measurements in conjunction with light-absorption spectra of carbonic anhydrase, with Co(II) substituted for Zn(II), have been utilized for band assignments of both the alkaline form of the enzyme and its anion complexes (*131*). It has been proposed that the active alkaline form of the Co(II) enzyme is a trigonally distorted tetrahedron (*131*).

[For other metalloproteins see also below, Dehydrogenases, Amine Oxidase, etc.].

C. Hemoproteins and Other Tetrapyrrole Proteins

1. *Myoglobin and Hemoglobin*

The oxygen-binding hemoproteins myoglobin and hemoglobin have been most thoroughly investigated with regard to their three-dimensional structure, based on X-ray analysis of their crystals (*132—135*). These data provide the basis for recent attempts to evaluate the optical activity of these hemoproteins in solution in a more quantitative manner than is usually possible (*136, 137*). Each polypeptide chain of these proteins contains iron-protoporphyrin IX(heme) as a conjugated prosthetic group which has symmetry and is optically inactive in isolated form.

a) Soret and Visible Regions. In the bound, native states optical activity is induced in the Soret and other visible-range heme transitions [see, for example, Refs. (*138—144, 10*), and Fig. 6]. Earlier interpretations of these Cotton effects included the assumption of strong interactions between the heme and the highly helical peptide environment (*84*). However, contributions to the rotatory strength in the Soret region from coupling of heme transitions with n-π^* and π-π^* transitions of the peptide amides have been considered to be negligibly small (*137*). Also, possible non-planarity of the heme or mixing of the heme π-π^* with either d-d transition of the iron or σ-σ^* transitions of the alkyl side chains have not been considered to account for the rotatory strength observed. The main contribution should come from coupling of the heme π-π^* electric transition moments with those of the allowed π-π^* transitions of the aromatic side chains of the globin. 12 such aromatic groups out of 23 in sperm whale myoglobin are considered to be sufficiently close for effective contribution to the rotational strengths by interaction with the porphyrin transitions. Known atomic coordinates from X-ray studies at 1.5 Å resolution have been used. The calculated net rotatory strength of the Soret band is 0.3 DBM, as compared to the experimental value of 0.5 ± 0.05 DBM for ferrimyoglobin. Analogous calculations for hemoglobin give 0.1 DBM per heme, as compared to 0.2 DBM obtained experimentally for human oxyhemoglobin (*137*). While these calculations are an important attempt to calculate rotatory strengths in a complex protein system, it should be remembered that they are based on combinations of a large number of data and include an inherent degree of uncertainty. Nevertheless, a similar approach may be applied to other hemoproteins or conjugated proteins, provided sufficient information on their three-dimensional structure will be available.

Some quantitative differences in the CD spectra in the Soret region of hemoglobin (*142, 145, 146*) have been attributed to the use of different slit widths (*147, 148*).

It has been suggested (*137*) that ligand binding at the heme iron (*149*) may change the orientation of the heme transition moments with consequent changes in the Soret CD bands, while the protein conformation would remain practically unchanged. On the other hand, variations in CD spectra upon oxygenation of hemoglobin have been attributed to changes in the tertiary structure of the protein (*150*).

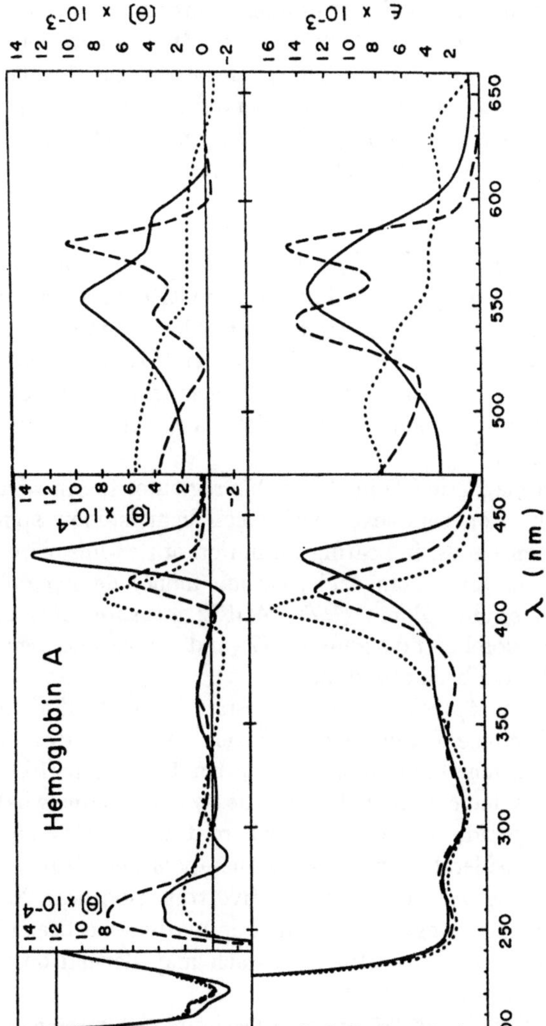

Fig. 6. Circular dichroism spectra and absorption spectra of native human hemoglobin A. The ellipticities (θ) and extinctions (ε) are on a molar heme basis. -----, oxygenated hemoglobin; ——, deoxygenated hemoglobin; ......, methemoglobin. [Reproduced from Ref. (*144*) where conditions are given]

Ligand-induced inversions of the negative Soret CD band of lamprey hemoglobin have been observed with alkyl isocyanides. The effect increased with the size of the hydrophobic residue. It has been suggested that a ligand-protein interaction displaces the heme chromophore whose Soret transitions interact with the protein environment according to the geometry created (*151, 152*). A different relative orientation between the heme and protein chromophores may also be the cause for the reversal of the sign of the Soret Cotton effect observed for the ferriheme protein of *Chironomus thummi* as compared to (sperm whale) ferrimyoglobin. More specifically, an opposite arrangement of coordinating ligands has been postulated in this case (*153*).

CD spectra have also been used to establish the presence of *p*H-dependent conformational isomers in ferrihemoglobin of *Chironomus thummi*, which had been postulated earlier from EPR data (*154*), and also of conformational isomers of horse ferrihemoglobin (*155*). Correlations between ellipticity changes and the Bohr effect (*154, 156*) and between the optical activity of heme transitions in hemoglobin derivatives and their absorption and magnetic properties have been suggested (*157*) [see also (*155, 152*)].

Natural and magnetic CD in the visible region of sperm whale myoglobin and of its iron complexes with small ligands show splitting of certain absorption bands, indicating formation of non-linear iron-ligand bonds; *e.g.*, in metmyoglobin azide, carbon monoxide-myoglobin and perhaps also in oxymyoglobin (*158*). MORD measurements on horse hemoglobin and dolphin myoglobin (*159*) and on various hemoglobin derivatives (*44*) have been reported.

Denaturation of *Aplysia* myoglobin results in practical abolishment of the positive Cotton effect in the Soret region, indicating disappearance of the asymmetric heme environment in the denatured state (*160*). Analogous observations have been made on the system sperm whale ferrimyoglobin-laurylpyridinium chloride, where the heme is apparently separated to a considerable extent from the apoprotein (*161*).

b) *Ultraviolet Region.* An optically active transition near 260 nm of both hemoglobin and myoglobin and their derivatives has been observed (*37, 162–165, 155*), the magnitude of which may depend on the spin state of the iron (*166*).

Changes in ellipticity at 285 nm of either horse or human hemoglobin vary linearly with the degree of saturation with oxygen. A correlation between these changes and an oxygen-linked conformational change in the protein has been suggested (*164, 162*).

Considerable interest has been focused on the far ultraviolet ORD and CD spectra of both hemoglobin and myoglobin and their derivatives, since these data have been used for estimation of the content of ordered and random structures in both hemoproteins and their respective apoproteins. Removal of the ferriheme from the protein, which is a reversible process, results in a significant change of the parameter b_0, of the rotation near 233 nm, or of the ellipticity near 220 nm, to more positive values, giving a lower calculated value for the apparent α-helix content in aqueous solution. The apoproteins of both sperm whale (*167, 168, 169*) and horse myoglobin (*170*) had an estimated helix content of about 50%, while the native hemoproteins had about 60–70% helix content, using the above criteria. Although the apparently low absolute values of helix content are uncertain[4]) for various possible reasons discussed above (see Optical Activity of Proteins) it, nevertheless, appears that the apoprotein has about 10–20% less in total possible α-helix content [see also Ref. (*172*)]. Similar results have been obtained with hemoglobin (*173, 62, 172*). Therefore, it appears that the prosthetic group heme stabilizes the secondary structure of the hemoprotein.[5]) A model system, consisting of poly-L-ornithine and ferriheme, which shows analogous effects, has been described above (*86*). In all the above estimates it has been assumed that heme (or aromatic group) transitions do not contribute significantly to the observed far ultraviolet rotations or ellipticities (*167, 138, 2, 149, 174, 60, 172*) [see, however, (*175, 163, 176, 177*)]. It has been noted above that no significant changes in the rotation at 233 nm were observed when ferriheme was added to helical poly-L-lysine of about 80% initial helix content, indicating that heme transitions do not substantially affect the changes observed at 233 nm (*86*). An interesting and pertinent effect of stabilization of peptide secondary structures by ferriheme has been reported on the binding of one molecule of ferriheme to one molecule of the apoprotein from horse hemoglobin: about one half of the helix content difference between the native protein and the apoprotein is restored in this case, as estimated by CD at 222 nm. It appears that even a polypeptide chain which does not carry ferriheme has its helix content increased cooperatively by binding of the heme to another chain (*62*).

The use of optical rotation for detection of subtle changes upon binding to a protein is also illustrated by a reported change of about 8% in the rotation value at 233 nm of human deoxyhemoglobin upon binding of oxygen or carbon monoxide. These rotational changes have been considered to be due to conformational changes of the protein. Myoglobin which consists of only one polypeptide chain per molecule or the isolated α- or β-chains of hemoglobin do not show the above

[4]) A value of about 78% α-helix is estimated from X-ray diffraction data on crystalline sperm whale ferrimyoglobin (*171*).

[5]) See also below, peroxidase and cytochrome b_5.

effects (*174, 177*). However, the full effect is observed with modified hemoglobins showing no heme-heme interaction (*177*). These above rotational changes have also been confirmed for hemoglobin recombined from deoxygenated chains (*145*), for hemoglobins S and F, and for normal adult hemoglobin dissociated into halves in 3 M sodium chloride (*178*) [see also Ref. (*177*)].

In contrast to the high α-helix contents of sperm whale myoglobin or horse hemoglobin, a respiratory hemoprotein from *Chironomus thummi* had about one third less α-helix content as estimated from its rotation at 233 nm (*153*); lamprey hemoglobin also indicated less α-helix content (*179, 144*).

In 0.742 M sodium chloride, the apparent α-helix content of human hemoglobin, as estimated from the ultraviolet optical activity, increased by 20—25% (of the maximum value), as compared to the value of hemoglobin in the absence of salt. This effect was attributed to a closer packing of the molecular units, leading to increased coupling between transition moments (*180*). CD and ORD measurements of human hemoglobin showed a decrease in the magnitudes of both the Soret and 260 nm bands with increasing sodium chloride concentration (*180*).

c) Subunits. CD measurements appear to be a sensitive tool to distinguish between isolated α- and β-chains of human hemoglobin (*145*). The CD spectra of these subunits differ both in the Soret (*142*) and visible regions (*147, 148, 144*), in all oxidation (see Fig. 7) and ligand states, indicating, at least in part, differences in the heme environment of the chains [for X-ray studies of horse oxyhemoglobin, see (*132*)]. Differences between α- and β-chains have also been recorded near 260 nm (*162*). In the range 300—400 nm (*146, 142*) α- and β-chains of human oxyhemoglobin A have been reported to exhibit different CD spectra, which are not additive with regard to oxyhemoglobin A. Additivity of CD band areas of α- and β-subunits in the Soret region, which give the values of reconstituted oxyhemoglobin, had previously been reported (*162, 147*). Significant non-additivity of CD bands has also been reported for subunits of deoxygenated human hemoglobin (*142, 144*). It has been concluded that certain structural features of the subunits change on recombination (*146*) [see also (*142, 148*)].

d) Abnormal Hemoglobins and Different Species. A comparison by ORD measurements of normal human hemoglobin A with human fetal hemoglobin F and with three abnormal hemoglobins (S, C, M_{Boston}) has shown that the Soret and visible-region Cotton effects of both oxy and deoxy derivatives of hemoglobins A, C, F and S are indistinguishable, while ferrihemoglobin M_{Boston} differs from hemoglobin A at *p*H 7. The M and F variants also differ from hemoglobin A in the region of the aromatic side-chain absorption (*178, 181*). By comparison of hemoglobin

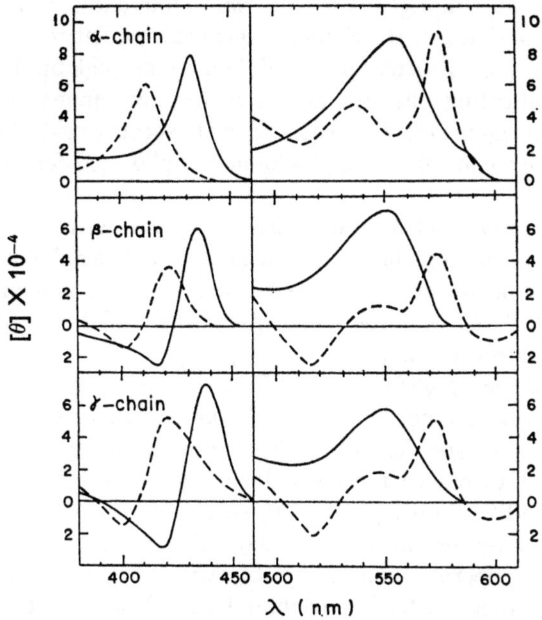

Fig. 7. Circular dichroism spectra of isolated α, β and γ chains (the latter from fetal hemoglobin). -----, oxygenated chains; ——, deoxygenated chains. [Reproduced from Ref. (*144*) where conditions are given]

Rainier ($\alpha_2\beta_2{}^{145\ \mathrm{Tyr}\rightarrow\mathrm{Cys}}$) with hemoglobin A, significant differences in the CD spectra of the deoxygenated forms and of the oxygenated β-chains have been observed (*182*). ORD measurements (500—600 nm) of sickle cell hemolysates have shown a large increase of the Cotton effects in the temperature range 0° to 38° in contrast to hemolysates from normal red cells. This difference indicates greater conformational changes with temperature in the sickle cell hemoglobin (*183*).

Some differences in secondary structure as measured by CD in the far ultraviolet regions have been recorded for various abnormal hemoglobins as compared to hemoglobin A (*181*).

CD measurements on both oxygenated and deoxygenated human and lamprey hemoglobins show considerable differences between the two species. In the Soret and near-ultraviolet regions, these variations have been attributed to different heme environments (*179, 144*).

Hemoglobin Bart's, which is composed of four identical subunits with no heme-heme interaction and Bohr effect, shows CD spectra (240—600 nm) different from those of human adult hemoglobin but similar to those of β-subunits of the latter (*184*).

e) Hemoglobin-Haptoglobin Complexes. Differences were observed in the optical activity in the Soret and near-ultraviolet regions between hemoglobin A (and its subunits) and human hemoglobin-haptoglobin complexes, indicating changes in the heme environment on binding (*185—188*). No significant differences in ESR spectra could be detected between the ferrihemoglobin-haptoglobin complex and ferrihemoglobin (*186*).

Some differences between measured CD data (near 250 nm) of a complex human haptoglobin 1-1 with horse ferrihemoglobin and data computed from measurements of the separate complex components were attributed to differences in the heme environment of the complex, while the secondary structure appeared to be preserved (*189*).

f) Artificial Hemoproteins and Dye Binding to Apoproteins. Complexes of globin (from bovine hemoglobin) with various porphyrins IX showed CD spectra in the Soret region which were qualitatively similar among themselves and to native ferrihemoglobin, but differed quantitatively in the Soret and 260 nm regions (*190*) [see (*144*)].

CD values observed at 222 nm and *p*H 9.2 for complexes of the apoprotein from sperm whale myoglobin with protoporphyrin IX are very similar to those of ferrimyoglobin. This indicates that the central iron is not required for the apparent stabilization of the secondary structure of the apoprotein by the conjugated molecule (*191*), as discussed above. Complexes of apohemoglobin (human) with either IX isomers of ferriprotoheme, ferrimesoheme, protoporphyrin, mesoporphyrin, copper-mesoporphyrin or manganese-protoporphyrin, all show similar ellipticities at 222 nm, again indicating that only the porphyrin ring affects the helix content (*192*). While a complex of apomyoglobin (sperm whale) with copper-protoporphyrin shows practically the same ultraviolet optical activity as ferrimyoglobin, the corresponding zinc complex or a derivative nitrated at the heme vinyl side chains gives different values which are close to those of the apoprotein (*193*).

Binding of related symmetrical dyes (naphthalene derivatives) to apohemoglobin resulted in different CD-Cotton effects in each case. As in the case of heme (see above), a coupled oscillator interaction with π-π^* transition moments of aromatic side chains of the protein was suggested, but no calculations were reported (*194*). The binding sites of the dyes used have not been identified and may include the heme "pocket" in some cases (*195*). The induced Cotton effects appear to be very sensitive to the local protein environment of the dye chromophores (*194*).

2. *Cytochrome c*

This oxidation-reduction protein has been investigated extensively and serves as an important example for structure-function relationships.

A considerable amount of work has been published on the optical activity of cytochrome *c* in both ferro and ferri states. As in other cases, earlier investigations were performed by ORD measurements.

a) Native Cytochrome c. Cotton effects in the near-ultraviolet, Soret and visible regions of monomeric horse heart cytochrome *c* showed large differences in both oxidation states, indicating changes in the heme environment (*196*) [these authors also cite earlier work in this field]; (*175, 197, 198*).

Similarly, a large Soret Cotton effect was reported for ferricytochrome *c* from horse heart and effects of opposite sign, for the reduced form. It was assumed that conformational changes associated with the observed changes in the ORD spectrum during the oxidation-reduction process were linked in some way to oxidative phosphorylation (*199*). More recent work mainly included CD measurements over a wide spectral range, which allowed for a better analysis of the complex data (see Fig. 8). These investigations generally reconfirmed the great differences between ferri and ferrocytochromes *c*, which were again attributed to significant changes in the dissymmetric environment of the heme (possible iron or ligand movement) and of the aromatic protein side chains (*201, 202*);

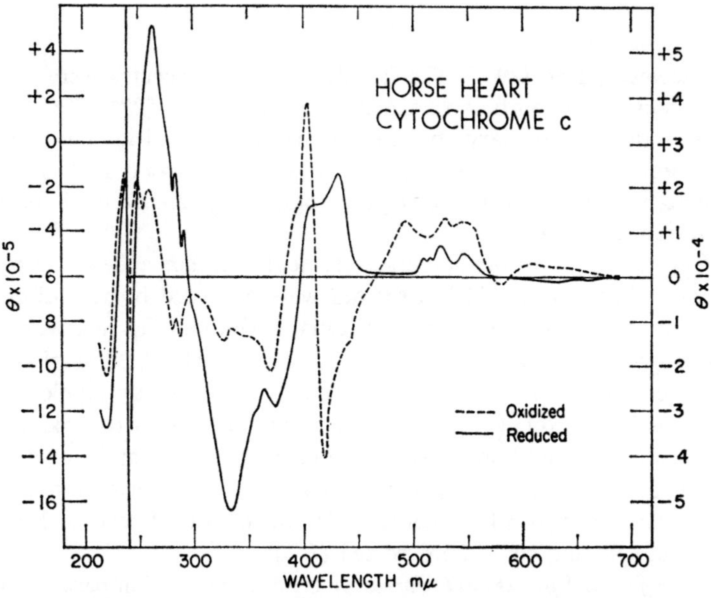

Fig. 8. Circular dichroism spectra of horse heart cytochrome *c* in 0.1 M potassium phosphate buffer, *p*H 7.0. Temperature, 25°. [Reproduced from Ref. (*200*)]

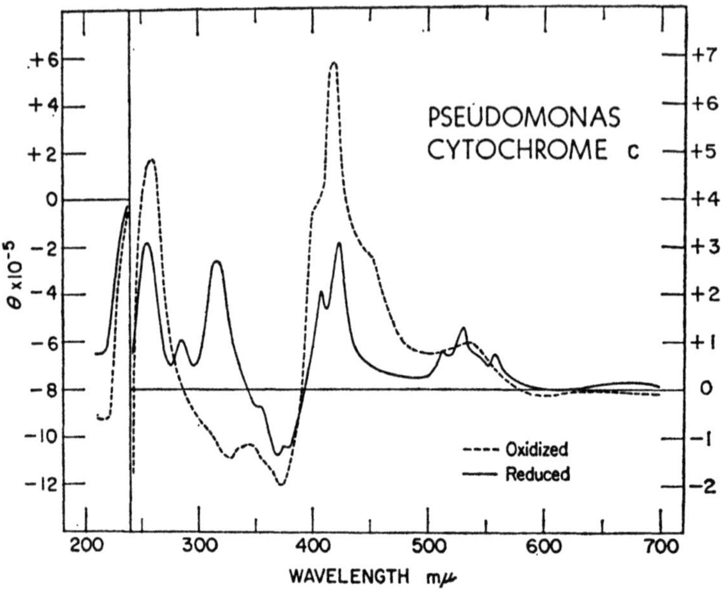

Fig. 9. Circular dichroism spectra of *Pseudomonas* cytochrome *c* in 0.1 M potassium phosphate buffer, *p*H 7.0. Temperature, 25°. [Reproduced from Ref. (*200*)]

[see however, Ref. (*200*)]. The far-ultraviolet CD spectra were reported to be similar for both ferri and ferrocytochrome *c* (*201, 203, 204*). It appears that the tertiary structure is mainly affected in the oxidation-reduction process. In any case, not more than about 10% α-helix is estimated in horse heart ferricytochrome *c* by X-ray data at 2.8 Å resolution (*205, 135*).

CD and light-absorption measurements on horse heart ferrocytochrome *c* in the range 570—1000 nm were analyzed by resolution into Gaussian bands. Using spectral criteria, including the relatively large anisotropy factors, transitions to the lowest crystal field singlet states were identified, some of which appeared to be magnetic dipole allowed *d-d* transitions [Ref. (*206*) and refs. cited therein]. Nitric oxide ferricytochrome *c* showed splitting of certain absorption bands, as revealed by CD spectra. This has been considered to be due to formation of a non-linear iron-ligand bond, in contrast to nitric oxide ferrocytochrome *c*, where no splitting could be detected (*158*).

b) Different Species and Forms of Cytochrome c. Comparative ORD data on various species and on modified derivatives of horse heart cytochrome *c* (*207, 208*) and data obtained at extreme *p*H values, at different salt concentrations and with denaturing agents (*209, 198*) were difficult to interpret. At least, ORD spectra could be considered to be more sensitive indicators for conformational changes than light-absorption spectra [*e.g.*, Ref. (*210*)]. A comparison of CD spectra (185—600 nm) of cytochromes *c* from horse, beef, tuna, turkey, pigeon and moth showed similarities in the oxidized state except for tuna cytochrome *c*. Larger differences were recorded for all these cytochromes in the reduced state, in particular in the visible region. A conformational homology among mammalian-type cytochromes *c* has been suggested on the basis of the CD data (*211*).

In the Soret region, the CD bands of ferricytochromes *c* from horse heart or other mammalian types and *Pseudomonas aeruginosa* (Figs. 8 and 9) appear to show bands of opposite sign. This reversal has been interpreted by a different arrangement of ligands at coordination positions 5 and 6 of the prosthetic group or by other environmental changes in its vicinity (*202, 200*). An analogous case involving ferrimyoglobin and the ferriheme protein of *Chironomus thummi* has been cited above.

ORD data indicated some differences between two isocytochromes *c* from yeast (*212*). The effects of *p*H and of the state of oxidation of cytochrome *c* from *Candida krusei* have been measured by CD (*213*).

The large dependence of the Soret-band Cotton effects on the state of oxidation observed in cytochromes *c* from various mammalian and other sources could not be seen with cytochrome c_4 from *Azotobacter vinelandii*. Its CD spectrum in the range 240—440 nm is rather similar to that of the heme undecapeptide from horse heart cytochrome *c* (see below). It appears that CD spectra of cytochromes are a better means of characterization and classification than light-absorption spectra (*214*).

CD measurements in the range 205—500 nm were reported for cytochromes from *Rhodospirillum rubrum* and *Rhodopseudomonas palustris* under various conditions, and the relatively high helix content of about 63% was estimated by standard methods for both ferricytochromes (*215*) [see, however, Ref. (*216*)]. ORD measurements on *Chromatium* cytochrome c_{552}, which contains two mesohemes and one flavin per molecule, showed complex behavior in the Soret region, indicating possible interaction between the hemes (*217*).

A comparison of CD spectra (200—600 nm) between bovine heart ferricytochrome *c* and ferricytochrome c_2 from *Rhodospirillum rubrum* showed marked differences, while the spectra were more similar in the reduced forms. These differences were attributed to subtle variations around the hemes rather than to differences in the polypeptide chain conformations (*216*). Optical activity studies of cytochromes c_3 showed similarities among three species of *Desulfovibrio* but revealed differences from mammalian cytochrome *c*. Rotatory strengths of the resolved CD bands indicated splitting of the Soret bands of both oxidized and reduced cytochrome c_3 (*218*). Four different forms of monomeric bovine heart cytochrome *c* have been separated by gel electrophoresis, and forms II to IV were considered to have conformations different from that of the native form I. ORD and CD measurements in the range 210 to 450 nm showed significant differences between the ferric states of forms I and II, on one hand, and form III, on the other, indicating conformational differences between them (*219*).

c) Cytochrome c Complexes and Effects of Medium and of Modifications. Cytochrome *c*-phospholipid complexes which are soluble in isooctane are considered to simulate native conditions in the cellular membrane environment. A comparison of the ORD spectra (220–600 nm) of this complex with those of cytochrome *c* showed similarities in both cases, indicating stabilization of the protein by the phospholipid against the apolar medium (*220*).

CD measurements were used as a sensitive indicator for monitoring conformational changes in horse heart ferricytochrome *c* effected by urea, *p*H, temperature and extrinsic ligands (*221*), by alcohols (*222*), or by chemical modifications (*223*).

d) Heme Peptides from Cytochrome c. Heme peptides derived from cytochrome *c* by proteolytic digestion were considered as model systems for the hemoprotein. The heme peptide obtained by peptic digestion showed Cotton effects in the Soret and visible regions very similar to those of native horse heart cytochrome *c* in both oxidized and reduced forms. This has been taken as an indication that the small heme peptide (11 amino acid residues) determines these Cotton effects (*224*). However, the complex Soret spectrum of the undecapeptide was considered to be due to polymerization, while the ORD spectra of the monomeric peptide are similar to those of denatured ferricytochrome *c* (*225*). ORD measurements of a heme octapeptide obtained from horse heart cytochrome *c* by tryptic digestion also revealed some differences in the Soret region when compared with the native hemoprotein. These ORD spectra appeared to be closer to those of myoglobin or hemoglobin, apparently because of the absence of one nitrogenous ligand in the non-aggregated heme peptide. In addition, large differences between native ferrocytochrome *c* and the reduced heme peptide complexes with imidazole or methionine derivatives were reported (*226*).

Further investigations of heme peptides from cytochrome *c* were carried out by both ORD and CD measurements (*227*, *163*, *228*, *202*). The aggregated hemepeptides were considered as model systems for interactions between hemes in biological systems. By analysis of light-absorption and CD data, exciton interactions between the Soret transitions in aggregated hemes were estimated. Under the conditions used, the aggregates of the undecapeptide appeared to be aligned head-to-tail (*227*) and in stacked or oblique conformations in the case of the octapeptide. In the latter case, maximum distances of 6–7 Å between hemes in aggregates were estimated, assuming mainly a dipole interaction potential between the heme transition moments (*228*). The heme undecapeptide was also used to identify observed transitions of hemoproteins in the ultraviolet region (*163*, *224*). Since this hemepeptide does not contain aromatic residues, disulfides and ionizable sulfhydryl groups, the CD bands observed in the range 240–360 nm have been attributed to the heme chromophore (*163*). Removal of iron from bovine heart cytochrome *c* or from its tryptic heme peptide, or of histidyl-18 from the latter, abolished optical activity around 260 nm (*216*). Thus, imidazole liganded to the iron may be essential for the optical activity of this transition (*166*). Previous ORD studies in the range 270–300 nm had shown Cotton effects which were dependent not only on the oxidation state of the iron of cytochrome *c*, but apparently also on its tryptophan content (*197*).

e) MORD and MCD. MORD spectra (300–600 nm) of beef heart cytochrome *c* exhibited a greater fine structure than corresponding ORD or light-absorption spectra measured at room temperature. While large differences between the oxidized and reduced cytochrome *c* were observed near 550 nm, there were only slight variations at the large negative Soret rotation (*229*). The effect near 550 nm was used for kinetic studies of oxidation-reduction reactions, which were interpreted as two-step processes of the electron transport (*230*). CD and MCD data of ferricytochrome *c* were also reported at longer wavelengths (500–850 nm) and at several *p*H values. Optical activity has been observed in the 695 nm band, which is considered to be a conformational indicator in ferricytochrome *c* (*231*). MCD spectra of horse heart cytochrome *c* were reported and analyzed (*232*, *158*, *233*). MORD studies on mammalian and yeast cytochromes *c* (*234*) showed some differences with results obtained in the Soret region (*229*, *230*). MORD data obtained for different species of ferrocytochrome *c* at various *p*H values were taken as possible indications for differences in the ligand fields at the central porphyrin iron (*235*, *44*).

3. *Cytochrome* P_{450}

This oxygen-activating hemoprotein from submicrosomal particles showed induced ORD Cotton effects in the Soret region in the oxidized and reduced carbon monoxide-complexed forms, while no Cotton effects were discernible for the reduced form of cytochrome P_{450} and for all states of cytochrome P_{420}. It was suggested that a conformational change occurs in cytochrome P_{450} during mixed function oxidation (*236*).

4. *Cytochrome* b_5

ORD and CD measurements (190—600 nm) of cytochrome b_5 from microsomes of pig liver were carried out under various conditions including denaturation. Reversibility of splitting and recombination of the protoheme was shown in the far ultraviolet region. Differences in the range 240—350 nm between native and apoproteins were attributed to interactions between the protoheme and aromatic residues (*237*). On the basis of CD, fluorescence and sedimentation experiments, it has been concluded that the environment of the single tryptophanyl residue of cytochrome b_5 remains apparently unchanged upon heme binding to the apoprotein, while some tightening of the tertiary structure and increase in the amount of secondary structure appeared to occur at *p*H 7.5 under these conditions (*238*). CD measurements in the aromatic amino acid region were used to follow conformational changes occurring on iodination of tyrosyl residues (*239*). CD spectra (200—600 nm) of larval cytochromes b_{555} and b_{563} in both oxidation states showed a variety of bands in all spectral regions (*240, 241*). Correlations between hemoprotein types and the sign of various Cotton effects were proposed (*240, 241*).

5. *Cytochrome Oxidase*

Some preliminary results obtained by ORD measurements on cytochrome oxidase have been reported simultaneously by several investigators [see Ref. (*242*), p. 503, Discussion on MORD and ORD of cytochrome oxidase]. Both in the Soret and visible regions, differences between the ORD spectra of the oxidized and reduced oxidase were recorded. These differences in the Soret Cotton effects, which were confirmed also by CD measurements, were attributed to changes in the heme environment caused by changes in the state of oxidation (*243—245*). Complex Cotton effects measured for ferrocytochrome oxidase were interpreted by resonance interactions of the Soret transition moments due to juxtaposition of the hemes (*243*). However, when the non-ionic detergent Emasol was used as a dispersing medium, resolution of the Soret bands showed the absence of ellipticity bands of opposite sign, with no evidence for heme-heme interactions (*246—247*). In the 600 nm region, oxidation state-dependent CD spectra were reported for beef heart cytochrome oxidase. These spectra were considered to originate from heme iron and copper chromophores interacting with the protein (*248*).

ORD measurements in the Soret region of the carbon monoxide derivative of cytochrome oxidase showed a single Cotton effect near 430 nm, which indicated that only cytochrome a_3 has an optically active transition, while the non-reacting cytochrome *a* subunit did not show measurable optical activity (*249*). By comparison of oxidation-reduction linked CD difference spectra of cytochrome oxidase in the presence of carbon monoxide and cyanide, further evidence in favor of two conformationally independent and distinct species (cytochromes *a* and a_3) was considered to have been obtained (*250*). However, CD studies of the carbon monoxide complexes of cytochrome *c* oxidase have led to the suggestion that the spectroscopic properties of this protein are determined by its state of aggregation (*251*).

ORD measurements on alkali-denatured cytochrome oxidase did not show Soret Cotton effects in the oxidized state, but a complex spectrum appeared at *p*H 11.6 in the reduced state, indicating heme-heme interactions. Polylysine-heme complexes were used as models (*252*). The optical activity of both oxidized and reduced heme *a* was measured in various solvents and under different conditions (*253, 254*).

6. *Catalase*

Early ORD measurements on beef liver catalase showed large Cotton effects in the Soret region, which disappeared upon denaturation (*255*). No evidence for previously reported (*256*) Cotton effects near 600 nm could be obtained (*255*). ORD and CD measurements on bovine liver catalase included also the liganded hemoprotein. Conformational changes in the protein, occurring upon ligand addition, were deduced from data in the ultraviolet region (*257*).

7. *Horse-radish Peroxidase*

Early measurements by ORD have been summarized (*10*). CD and light-absorption spectra of ferri-, cyanoferri-, ferro-, and carbon monoxide-ferro derivatives of HRP have been compared with those of analogous derivatives of sperm whale myoglobin. In all cases, resolution into Gaussian curves has been attempted. In contrast to myoglobin, which also contains one molecule of the same prosthetic group per molecule of protein, both the reduced HRP and its carbon monoxide derivative yield CD spectra which are very different from each other, indicating local conformational changes occurring near the heme upon reduction of HRP. Besides transition dipole coupling between the heme and aromatic side chains of the protein, inherent dissymmetry, produced by some deformation of the bound heme, has been considered as a possible source for generation of optical activity (*141*) [see, however, Ref. (*137*)].

The formation of HRP-substrate compounds shifts the CD and light-absorption bands in the Soret region in a similar fashion. Differences observed in the CD near 280 nm between the enzyme and the enzyme-substrate complex could reflect changes in orientation of aromatic amino acid residues upon binding, provided the heme does not contribute to the CD in this region (*258*).

Two peroxidase isoenzymes from horse-radish roots and their apoenzymes were compared structurally by measurement of their respective CD spectra over the range 180—700 nm. It was concluded that the active sites are similar in both isoenzymes, despite some observed spectral differences (*259*).

According to ORD measurements of ferric HRP and its apoprotein, splitting of the heme appears to cause a reduction of the estimated α-helix content (*260, 259*). CD spectra of Japanese-radish peroxidase *a* have been interpreted as reduction of about 35% (upper limit) in the α-helix content upon removal of the heme with possible formation of β-structure. Differences between the oxidized and reduced forms of this peroxidase have been recorded (*261*) [see above, Myoglobin and Hemoglobin, ultraviolet region].

8. *Phytochrome*

This plant chromoprotein has two forms (P_R, P_{FR}) interconvertible by light, and contains a removable conjugated chromophore which is considered to be a tetrapyrrole [see Ref. (*262*), and refs. cited therein]. CD measurements up to 800 nm indicate a change from a negative band at 660 nm for P_R to a positive band at 725 nm for P_{FR} upon irradiation by red light (*263*), while ORD and CD measurements at 10 °C and in the range 200—500 nm show little α-helix content and large differences between the two forms P_R and P_{FR} (*262*). At low temperatures (—70° to +4 °C), four different and interconvertible forms of phytochrome in 66% glycerol have been characterized by light absorption and CD. Correlations between the spectral intensities and possible conformations of the prosthetic group have been based on Hückel molecular orbital calculations [(*264*); and refs. cited therein].

9. *Phycocyanin*

Multiple Cotton effects in the visible region were observed for the algal chromoprotein phycocyanin (*265, 266*). Interpretations included inherent dissymmetry of the tetrapyrrole chromophore (*267*), possibly induced by binding to the protein, interactions between chromophores of different monomeric units or between these chromophores and protein transitions (*265, 266*). The chemistry and mode of attachment of the conjugated molecules to the protein will have to be elucidated in more detail in order to enable a better evaluation of the origin of the optical activity observed.

10. *Phycoerythrin*

Different CD spectra were reported for "single"- and "double-peaked" phycoerythrin obtained from fresh and aged blue-green algae, respectively (*268*). For single-peaked phycoerythrin (absorption at 560 nm), two major CD bands of opposite sign were recorded above and below 560 nm, indicating chromophore interaction (splitting), while two positive CD bands were measured in the absorption bands of the double-peaked phycoerythrin. Added *p*-mercuriphenyl sulfonic acid converted the two proximate CD bands of opposite sign of the single-peaked material to a complex positive band, indicating loss of interaction between the chromophores under these conditions (*268*).

11. Bacteriochlorophyll Protein

CD measurements at 77 °K of a bacteriochlorophyll protein from *Chloropseudomonas ethylica* indicated exciton splitting near 800 and 600 nm of bacteriochlorophyll *a* molecules. Distances of 12—15 Å between the interacting molecules in a protein subunit were estimated (*269*).

D. Dehydrogenases

1. Flavoproteins

Cytochrome b_2 (yeast L-lactate dehydrogenase), which carries both protoheme and FMN as prosthetic groups, has been studied by both fluorescence and ORD-CD measurements (*270—273*). Complex data obtained also for derivatives of cytochrome b_2 have been interpreted as interactions between the flavin and heme (*270—272a*). The Soret band of the heme apoprotein is inverted in sign as compared to the enzyme in which the flavin is present (*270, 273*). This has been attributed to a conformational difference in the protein in both cases (*273*). Preliminary MCD data obtained at low temperatures on oxidized cytochrome b_2 have been reported (*274*). An extensive analysis of light-absorption, CD, and MCD spectra of various free and protein-bound flavins has been described (*275—277*). CD spectra indicate that the environments of the flavins of several flavoproteins are similar in all oxidation states (*276*). CD spectra observed in the visible range of two flavoproteins (*Clostridium pasteurianum* and *Desulfovibrio vulgaris*) have helped to distinguish two subclasses of dehydrogenases which differ in riboflavin binding (*278*). Far ultraviolet CD spectra indicate that the coenzyme of a flavoprotein from *Azotobacter vinelandii* can be removed reversibly from the apoprotein (*277*).

The optical activity of xanthine oxidase has been described above (see Metalloproteins).

Ultraviolet light-absorption and CD spectra (190—420 nm) of reduced and oxidized mammalian lipoamide dehydrogenase have been measured and analyzed (*279, 280*). Changes observed in its CD spectra upon binding of FAD suggest conformational and environmental differences of the coenzyme in the free and bound states. The CD spectra of the reduced state depend on the reducing substrate, again indicating conformational and environmental differences between these reduced states (*280*).

ORD and CD of D-amino acid oxidase from pig kidney showed visible-region Cotton effects, which were dependent on the oxidation state of the coenzyme FAD (*281*). An inversion in the sign of the dichroic band near 372 nm was observed when free FAD was bound to the apoenzyme of D-amino acid oxidase, indicating a conformational change in FAD upon binding (*282*), while optical activity was induced in substrates or inhibitors of this oxidase [Ref. (*283*) and refs. cited therein]. Cotton effects of L-amino acid oxidase were also observed in the absorption region of FAD (*284*).

Native dihydroorotate dehydrogenase, which contains FAD, FMN and iron, showed an ORD Cotton effect near 420 nm, which changed upon reduction by the substrate. An absorption band has been observed in this region for the flavin-free iron protein. It has been suggested that the iron of the enzyme is reduced by the substrate (*285*).

2. *Pyridine-Nucleotide Dehydrogenases*

Earlier work, in particular on ORD of liver alcohol dehydrogenase, has been summarized (*10*). ORD of rabbit glyceraldehyde-3-phosphate dehydrogenase has been studied as a function of coenzyme binding. Changes in ORD values, observed upon removal of the coenzyme, could largely be restored upon addition of only one equivalent of coenzyme to the apoenzyme (*286*). Different results were obtained by other investigators who also reported CD spectra of both apo- and holoenzymes. A broad, positive CD band near 350 nm, and changes in the aromatic region were observed upon addition of NAD^+ to the apoenzyme (*287, 288*). Changes in the ORD parameter b_0 of glutamate dehydrogenase upon binding of NADH, both in the presence or absence of the inhibitor GTP, have been interpreted as conformational changes of the enzyme (*289, 290*). ORD and CD were used as a measure of conformational changes in yeast glyceraldehyde-3-phosphate dehydrogenase upon binding of the coenzyme NAD, and were correlated with other physical properties such as the sedimentation coefficient. A transition between 20°—40 °C from non-cooperative to cooperative binding of NAD by the apoenzyme was evaluated (*291, 292*). In contrast to pig skeletal muscle lactate dehydrogenase, no changes in the ultraviolet ORD spectra were recorded upon binding of NADH or oxamate to porcine heart lactate dehydrogenase (*293*).

The complex between beef liver glutamate dehydrogenase and NADH showed a positive CD band near 350 nm, in contrast to free NADH (*294,*

295). Addition of L-glutamate caused an inversion of the band. Analogous results were obtained with NADPH (*296*). Again, conformational and environmental changes probably are the cause of these effects, while no detailed explanation of these phenomena is as yet possible. CD spectra also suggested the existence of a second binding site for NADH in the presence of GTP or zinc ions (*294*, *297*). The appearance of a negative CD band near 260 nm in the complex of the enzyme with ADP was reported (*298*). Exchanging the zinc of horse liver alcohol dehydrogenase for cadmium or cobalt resulted in characteristic enzymatic and optical properties of the enzyme, which were also reflected in changes of CD spectra of aromatic residues and of NADH or 1,10-phenanthroline bound to the dehydrogenase (*299*). CD was also used as an indicator for binding studies of the symmetrical dye bis (1-anilino-8-naphthalene sulfonate) to the multiple forms of lactic dehydrogenase (*300*). The interaction of streptococcal dihydrofolate reductase with substrates, coenzymes and inhibitors was followed by CD in the ultraviolet region (*301*).

Optical properties, including optical activity, of the NAD coenzyme have been summarized (12). Adenosine-5′-nicotinate (ARN), in which the stacked bases are on the same side of the ribose, has been used as a model system for demonstration of coupling effects between proximate and strong electric transitions (*12*). Changing the equilibrium between stacked and non-interacting conformations by variations of *p*H, solvent, or temperature, reciprocal relationships between the rotatory strengths of the relevant CD bands of opposite sign could be demonstrated (see Fig. 10, NADH). Analogous data were also obtained for FAD.

By resolution of both light-absorption and CD bands of ARN and FAD and by the use of both experimental and theoretical values for transition moments, conformations of the relevant molecules were deduced. However, because of difficulties in accurate assignment and determination of the various transition moments and because of the limitations inherent in calculations using the coupled oscillator model and dipole-dipole interaction potentials, only a qualitative picture of the conformations involved could be expected (*12*). Generally, reciprocal relations between known transitions were taken as evidence for juxtaposition of the corresponding chromophores (*12*). The optical data obtained for coenzymes or their model compounds may be utilized in favorable cases for estimating structural features in the bound state, in which the interactions with the protein environment will also have to be evaluated.

Recently optical properties, including CD and ultraviolet spectroscopy, have also been determined for "abbreviated" synthetic NAD^+ analogues in order to study the intramolecular interaction between the quaternized nicotinamide and the adenine in aqueous solution. This

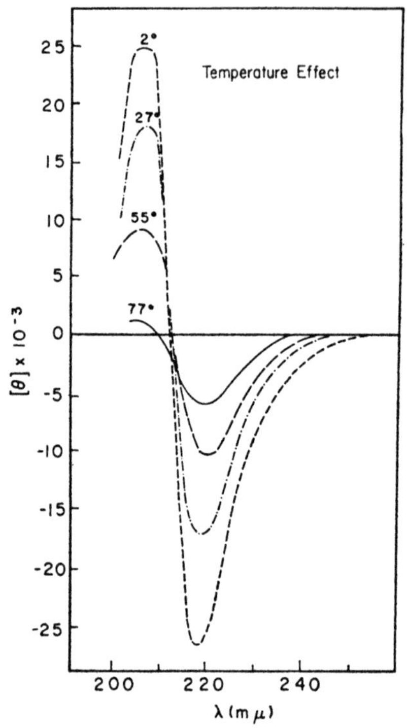

Fig. 10. Circular dichroism curves of α-NADH at various temperatures in 1 M NaF at *p*H 7. Reciprocal relations are observed between a reduced nicotinamide band near 215 nm and an adenine band near 210 nm [Reproduced from Ref. (*302*)]

model, which did not include one ribose and diphosphate groups, appeared to show larger interaction between the base rings than NAD^+ (*303*).

E. Proteins Combined with Different Small Molecules

1. Serum Albumin

A considerable amount of work has been devoted in recent years to the optical properties of molecules conjugated with serum albumin, which has physiologically important functions as a protein transporting various substances in the blood.

Mercurated fluorescein derivatives, which were covalently linked to the sulfhydryl group of bovine serum albumin (BSA), showed induced Cotton effects in the region of the dye absorption, which disappeared in 8 M urea (*304*). The dependence on *p*H of the Cotton effects indicated conformational changes in the protein (*304*), which were correlated with two conformational transitions occurring at *p*H 4—5 and 6—8, respectively (*305*). These apparent changes in the protein were not reflected in the rotation at 233 nm or in the b_0 value (*304*). Large optical activity appeared near the isoionic point of *p*H 5, where the albumin is in a contracted state (*305, 304*). As in many other cases cited in this review, it appears that changes in the tertiary structure of a protein can be followed by measurement of the optical activity of the conjugated molecule. On the other hand, the CD spectrum (300—500 nm) of a complex of BSA with pyridoxal-5-phosphate was independent of *p*H in the range *p*H 5 to 8 (*305*). A different binding site is apparently involved,

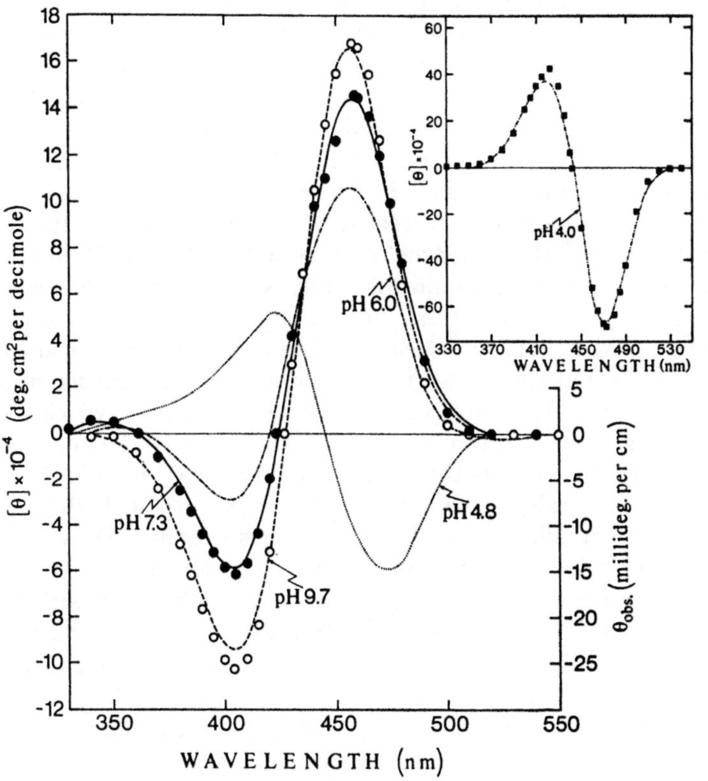

Fig. 11. CD spectra of the complex bilirubin-(charcoal-treated) human serum albumin at different *p*H values. Temp., 27.0 ± 0.5 °C. ■, measured at *p*H 4.00 ± 0.02 (inset); ●, measured at *p*H 7.3 ± 0.1; ○, measured at *p*H 9.7 ± 0.3. Drawn curves at *p*H 4.0, 7.3 and 9.7- data calculated by computer from experimental data (points). Resolution into two Gaussian curves at *p*H 4.0 and 9.7, and into three Gaussians at *p*H 7.3. Drawn curves at *p*H 4.8 and 6.0 are based on experimental data only. [Reproduced from Ref. (*308*) where conditions are given]

and it seems that no structural changes can be monitored in this case. Very large rotatory strengths of optically active transitions in the visible range (400—500 nm) of the physiologically important complex bilirubin-serum albumin were measured, in particular at *p*H 5, for the complex bilirubin-BSA (*306, 307*) and at *p*H 4 for bilirubin-human serum albumin (HSA) (*308*). The proximate CD bands were of opposite sign (see Fig. 11).

In the absence of the proteins and under similar conditions, no significant optical activity could be measured for bilirubin in solution (*307*). At certain *p*H values, added electrolytes or fatty acids had considerable effects on the magnitude or the sign of the dichroic bands (*309—311*). A substantial part of the observed Cotton effects was attributed to exciton coupling between electric transition dipole moments of the dipyrromethene chromophores of dissymmetrically bound bilirubin; however, other contributions, such as interactions with the asymmetric protein environment or inherent dissymmetry resulting from skewed chromophore conformations, may also be present (*307, 312, 308*). With some assumptions made, both distances and relative positions in space of the relevant transition dipoles of bilirubin were estimated for various cases in which the differences in the magnitude and sign of the optical activity of the exciton bands were attributed to variations in the internal rotation around the center methylene bridge of bound bilirubin in the presence of excess serum albumin (*308*) [see some analogies with 1,1'-bianthryls (*313*)]. Thus, measurement of optical activity may constitute a very sensitive probe of the conformation of both the protein and of the bound molecule. This is also apparent from the diverse results obtained with serum albumins from different mammalian species, which indicate differences in the microenvironment of the respective binding sites for bilirubin (*314*). The optical activity of human blood serum in the visible region shows analogous features to those measured with the complex prepared from bilirubin and HSA (*315*). This suggests CD as a possible method for determination of bilirubin in the serum. Similar exciton-type CD spectra are also observed upon interaction of bilirubin with sodium deoxycholate (*316*). Complexes of HSA with biliverdin, which has a methene bridge in its center, do not show substantial exciton splitting in visible-region light-absorption or CD spectra upon binding, in contrast to analogous complexes involving bilirubin, which has an isolating and rotating methylene bridge in its center (*317*).

Large Cotton effects were reported to be shown by the bilirubin-HSA system at excess bilirubin, where microcrystals formed on the asymmetric protein were considered to generate optical activity (*318*). Analogous arrays may originate from quartz surfaces (*307*).

The interaction with HSA of drugs such as phenylbutazone and its analogues (*319*), dicoumarol (*320*), warfarin, 4-hydroxy-coumarin or salicyclic acid (*321*), induces the formation of CD bands in these drugs in the ultraviolet region. The rigidity of the ligand-macromolecule complex has been considered to be important for generation of optical activity (*319*). Some differences in the CD spectra of drugs with serum albumins of various species have been recorded, in particular for rabbit albumin (*319, 322*).

CD measurements in the range 250—300 nm have been used for evaluation of binding constants of L-tryptophan to HSA, since additional dichroic bands are produced at a molar excess of the amino acid (*323*).

The symmetrical dye 1-anilino-8-naphthalene sulfonate acquires optical activity upon binding to BSA (*324*). With a small average number of bound dye molecules per molecule of protein, two adjacent CD bands of opposite sign are recorded which, however, are not considered to result from dipole-dipole coupling between dye molecules. The variation of the CD spectra with the average number of bound dye molecules appears to be sufficiently sensitive to distinguish among small intrinsic differences in the affinity of the binding sites (*324*).

CD spectra in the range 300—600 nm were used to locate the binding site of Cu(II) to BSA, by comparing these spectra with those of Cu(II) bound to some BSA fragments (*325*). CD spectra in the above range of complexes between cupric or nickel ions and model peptides also provided a basis for interpretation of the binding of these metal ions to various proteins (*e.g.*, BSA, conalbumin, α-chymotrypsin) in aqueous alkaline solution (*326*). In all cases, the first metal ion appears to chelate at the terminal amino group of the protein, while in the case of BSA, the second or third ion seems to interact with the sulfhydryl group of this protein (*326*).

2. *Visual Pigments*

A recent review on ORD and CD of visual pigments, which includes literature up to about 1969 (*327*), is available. Earlier work on the optical activity of visual pigments is also summarized in Ref. (*328*) [see also (*329*)], which describes the disappearance of a CD band in the visible region upon bleaching by white light, using visual pigments of various species in solution. ORD and CD measurements in the far ultraviolet region of bovine rhodopsin in aqueous detergent solutions indicate about 60% α-helix content, as calculated by conventional methods, while bleaching results in an apparent loss of some of the helix content (*330*) [see also Refs. (*331—333, 328*)]. However, the important question of whether or not a conformational change in the opsin occurs on photoisomerization of rhodopsin, which may cause triggering of electrical impulses (*331*), remains unanswered, since no conformational change in the ultraviolet region has been observed with outer segments (*330*) [see

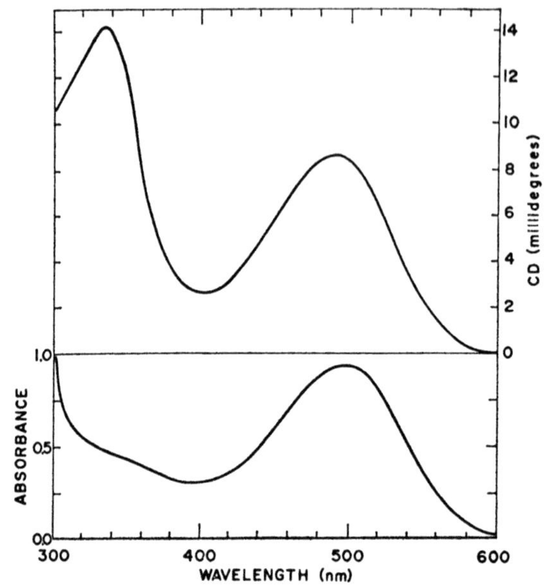

Fig. 12. Absorption and circular dichroism spectra of rhodopsin in digitonin solution at *p*H 7.7, 3°. The path length was 1 cm and the concentration of rhodopsin was 2,23 $\times 10^{-5}$ M. [Reproduced from Ref. (*335*)]

(*334, 335*)]. CD spectra at 3 °C of physiologically important intermediates in the bleaching of bovine rhodopsin have been recorded [Ref. (*335*) and refs. cited therein]. Because of the observed similarity of the spectra in the visible region (see Fig. 12) between rhodopsin and metarhodopsins (both in digitonin solution and in sonicated particles of rod outer segments), it has been suggested that a significant part of the local environment of bound retinal is conserved in the transitions from rhodopsin to metarhodopsins I and II (*335*).

The generation of CD bands in the visible region of solubilized cattle or squid rhodopsin and isorhodopsin was attributed either to internal twisting of the 11-*cis* retinal, causing inherent dissymmetry, or to its coupling with protein chromophores (*328, 333, 336*). Schiff's bases prepared from either *cis* or *trans* retinaldehyde and various aromatic amines showed optical activity in organic solvents. Rotatory strengths were calculated on the basis of a coupled oscillator mechanism. It has also been suggested that, upon binding to the opsin, similar coupling of the conjugated molecule with protein transitions generates the optical activity, rather than internal twisting of the retinal chromophore (*337*).

CD spectra measured on bovine rhodopsin in the near-ultraviolet and visible regions were used as additional evidence for recognition of the band near 345 nm (β-band) as part of the spectrum of unbleached rhodopsin (*338*). CD data of native and illuminated bovine visual pigment$_{500}$, obtained by illumination at low temperatures (77 °K), indicated complex, unexplained phenomena of two interconvertible states (*339*). As solubilized rhodopsin may not be a good model, CD spectra of bovine rhodopsin, associated with membrane particles and prepared by sonic disruption, have been measured. While the absorption of these preparations was similar to that of purified rhodopsin, there were quantitative differences in the visible-range CD spectra. It was suggested that phospholipids are involved in stabilization of conformations of both the opsin and the asymmetrically bound chromophore (*340*).

3. *Neurophysin*

Changes in the near-ultraviolet CD of native or modified bovine neurophysins occurring upon binding of hormones or small synthetic peptides were used for structural interpretations (*341—343*).

4. *Avidin*

Optical rotation in the range 260—300 nm was used to demonstrate protection against heat denaturation of avidin from egg white upon complex formation with biotin at *p*H 9 (*344*).

5. *Immunoglobulins*

CD has been considered as a sensitive method for comparison of combining sites of antibodies with nitrophenylated haptens (*345*). With antinitrophenyl antibody fragments of rabbit γ-immunoglobins, no significant differences in the far ultraviolet CD spectra with the antibody-hapten complex have been observed (*346*).

Ultraviolet CD spectra reveal significant differences in three fractionated and homogeneous rabbit antibodies against the capsular polysaccharide of Type III. Attributing these observed variations to changes at or near the binding site of the hexasaccharide, it has been suggested that the structures of these sites differ, although the antibodies are of similar specificity and affinity. This degeneracy may have distinct biological advantages (*347*). The specific binding of 2,4-dinitrophenyllysine to a mouse myeloma protein has been followed by CD titration in bands appearing in the complex. At least part of the induced optical activity in the range 300 to 600 nm has been attributed to coupling between electric transition moments of the dinitrophenyl group and of aromatic protein chromophores at the antibody combining site (*348*).

6. *Lipoproteins*

Far-ultraviolet CD studies of high-density lipoproteins from human serum indicated a high α-helix content for both lipoprotein and apoprotein. Denaturation experiments, which were followed by CD measurements, suggested stabilization of the protein by lipids through hydrophobic interactions (*349*) [see also Ref. (*350*)]. CD measurements in the far ultraviolet region, combined with infrared measurements, indicated the occurrence of β-structure in β-lipoprotein (*351*).

7. Actin

CD spectra in the near- and far-ultraviolet regions of the muscle proteins G- and F-actin, in combination with either ADP or 6-mercaptopurine nucleotide, and the effect of polymerization on the bound nucleotide were measured (*352*). Far-ultraviolet ORD and CD spectra of G-actin, measured under various conditions, were reported (*353*).

8. Phosphorylase

Muscle phosphorylase b consists of two protein subunits, each of which binds one molecule of pyridoxal phosphate. Optical activity can be demonstrated in the absorption band (near 333 nm) of pyridoxal phosphate by CD measurements, while in ORD spectra this Cotton effect is obscured by the protein (*354*). As measured by rotation in the ultraviolet region, the pyridoxal phosphate has little effect on the ordered structures of the protein (*355*).

9. Amine Oxidase

This enzyme from *Aspergillus niger* has been characterized to be copper-pyridoxal phosphate dependent. A CD band near 330 nm was attributed to the bound pyridoxal phosphate; and a band near 470 nm, to its Schiff's base with the substrate (*356*).

10. Aspartic Amino Transferase

ORD measurements in the range 300—600 nm of the native enzyme showed Cotton effects due to the coenzymes (pyridoxal and pyridoxamine phosphate). These effects were absent in the apoenzyme, in the coenzymes alone, and at high substrate concentrations (*357*). CD data obtained for this enzyme were transformed to ORD curves, which were in good agreement with ORD data measured previously (*358*).

11. L-*Glutamic Acid Decarboxylase*

Induced CD bands were observed for L-glutamic acid decarboxylase in the bands of the pyridoxal phosphate coenzyme near 400 nm [(*359*) and refs. cited therein].

12. Creatine Phosphokinase

ORD and CD spectra of creatine phosphokinase and of its complexes with adenine nucleotides, both in the absence and presence of metal ions and substrates, were measured in the range 190—600 nm. The large Cotton effects measured in the absorption bands of the bound chromophores were considered to be due to dipole coupling between the adenine transition moments and those of protein groups. Theoretical estimates, based on measured dipole and rotatory strengths and on the expected geometry of interacting groups, indicated a main contribution to the rotatory strength of the adenine from interactions between ADP and a tryptophan residue (*360*).

13. Pyruvate Decarboxylase

Conformational analysis by ORD and CD of yeast pyruvate decarboxylase dependent on thiamine pyrophosphate showed large changes in the aromatic region (282, 288 nm) and in the absorption band of thiazolium ion (near 262 nm), indicating interactions between the thiamine pyrophosphate and tryptophan (*361*).

14. Aspartate Transcarbamylase

By using CD difference spectroscopy in the range 250—310 nm, deductions about the role of tyrosyl and tryptophanyl residues upon ligand binding or subunit interactions were made (*362*).

15. Lysozyme

Specific binding of the competitive inhibitor N-acetyl-D-glucosamine to lysozyme modified the CD spectrum in the range 240—300 nm, which indicates changes in the orientation of the aromatic amino acid residues at the binding site (*363*). Association constants were determined from the CD data (*364*).

16. Ribonuclease

ORD and CD data indicated conformational changes in the interaction between ribonucleases and various nucleotides (*365—368*). CD studies of a complex of T_1-ribonuclease with guanylic acid and its analogues showed large dichroic bands of opposite sign at 250 and 280 nm, respectively, which were interpreted to be due to dipole coupling between purine chromophores and those of aromatic protein groups near the active site (*369*).

Modified azo derivatives of ribonuclease showed induced CD bands. Other azoproteins also exhibited characteristic CD spectra in the range 300—600 nm (*370*), while different chemical modifications have also been used to monitor structural and conformational characteristics of proteins by measurement of optical activity [see refs. cited in Ref. (*370*) and Ref. (*371*)].

F. Proteins Combined with Other Macromolecules

1. RNA-Synthetase

CD studies in the far-ultraviolet region of tyrosyl-RNA synthetase from baker's yeast indicated changes in ordered structures, provided the active enzyme was incubated with its specific *t*-RNA. The conformational changes observed may be related to the enzyme function (*372*).

2. Nucleohistones

CD spectra (200—350 nm) of calf thymus chromatin were compared with those of pure DNA in solution. Some of the differences observed (above approximately 250 nm) were attributed to conformational changes in the bound DNA, in which a large tilting of the bases may occur [(*373*) and refs. cited therein]. Previously, similar conclusions had been reached on the basis of ORD measurements on native nucleohistone and on nucleohistone dissociated either fully or partially into DNA and histone [(*374*); see also (*375—379*)]. CD studies on the state of DNA in native and partially deproteinized gander erythrocyte chromatin and of complexes of ethidium bromide with these systems were reported (*380*).

Another report indicates essentially no changes in the CD spectrum in the range 240—300 nm of calf thymus nucleohistone IV—DNA complexes, as compared to DNA alone; however, the far-ultraviolet CD spectrum of histone IV changes considerably in the complex, but does not conform with known structures (*381*). Other results obtained also by CD measurements indicate, however, that histone IV plays an important role in determination of the DNA conformation in chromatin and that it may act as a general repressor of RNA synthesis (*382*). Calcium nucleohistone IV is considered to be a structural model for chromatin, as appears from CD measurements (*383*). The DNA of intact nuclei shows a different CD spectrum (non-conservative), as compared with lysed nuclei or chromatin, indicating a specific geometry of DNA packing in the nucleus (*382*). In studies involving sedimentation, electron microscopy and CD of sheared native calf thymus chromatin, correlations between the state of aggregation and changes in the CD spectra have been confirmed (*384*). CD data obtained in the region of 275 nm indicate that in 2 M NaCl the conformation of DNA in soluble calf thymus histone is identical with that of native DNA because of electrostatic considerations (*385*).

A conformational assessment of the DNA-histone complex and of its components at various *p*H values, ionic strengths and temperatures indicates that both components are mutually stabilized in the complex against denaturation (*386*). A recent CD study of histone-DNA complexes shows that histone I (*f*-1) is largely in the extended conformation, while histone IV (*f*2*a*1) may have some α-helix content [(*387*) and refs. cited therein; (*388*)]. Specific effects of electrolytes and dioxane on the histone *f*-1-DNA complexes, as measured by CD, have been reported (*389*) [see also (*390*)]. DNA complexed with histone fractions *f*-2 and *f*-5, isolated from avian erythrocytes, indicates conformational changes in DNA-*f*-2 but none in DNA-*f*-5, as measured by ORD (*391*).

It appears that, although optical activity is a very sensitive and unique tool for detection of conformational changes, no quantitative relationships are so far available for complicated systems such as nucleohistones.

3. Tobacco Mosaic Virus

On the basis of ORD data it has been concluded that RNA in TMV is in a rigid single-stranded conformation similar to dinucleoside phosphates (*392*).

By combining the results of calorimetric, hydrodynamic and ORD-CD measurements on TMV-RNA obtained at different ionic strength it has been suggested that conformational changes, most likely as a result of base pairing, are responsible for heat effects observed (*393*).

VI. Summary and Conclusions

The above survey of recent work on the optical activity of conjugated protein systems demonstrates the increasing use of ORD and CD as very sensitive methods for their characterization and for elucidation of structural and conformational properties of these and other macromolecular systems in solution.

For conjugated proteins it seems convenient to divide the spectral range usually covered in ORD and CD measurements (190—700 nm) into 3 main regions:

1. The far ultraviolet region of peptide amide absorption.

2. The region of absorption of aromatic amino acid residues (about 250 to over 300 nm).

3. The near-ultraviolet and visible regions, where transitions of many of the bound non-protein molecules will be measured, while protein transitions will not contribute practically in CD measurements. Evidently, various molecules conjugated to proteins may also directly affect the ORD and CD spectra of the proteins in regions (1) and (2).

Generally, the following useful information may be obtained from measurement of optical activity of conjugated protein systems:

a) Symmetry or asymmetry of the protein environment of a bound molecule.

b) Conformational changes under various conditions in both three-dimensional and secondary structures of proteins.

c) Conformational characteristics of certain bound molecules under various conditions. These include internal conformational changes in the bound molecules, in certain cases, and changes in their positions relative to an asymmetric protein environment.

d) In favorable cases, the geometry of the interacting groups, which includes the handedness of an asymmetric center, may be determined.

e) Identification of transitions. This includes resolution of exciton coupling effects, characterization of transitions which cannot be recognized by other spectroscopic methods, or location of (aromatic) amino acid residues proximate to the bound molecules.

f) Titrations between the protein and the bound molecule for independent evaluation of binding data.

It follows that the measurement of optical activity of conjugated protein systems may be useful in following various important molecular processes such as allosteric transitions, aggregation, denaturation and various structure-function relationships. Because of the sensitivity of these methods, originating frequently from the dependence on angles and distances between the interacting chromophores, analytical differentiation among closely related protein systems may also be possible. Of major importance appears to be the possibility of directly following three-dimensional characteristics of the protein carrier in solution by using non-proteinous parts as conformational marker molecules. This applies also to artificial protein complexes. For most conjugated protein systems, CD will be the preferred method of measurement. It should always be coupled with corresponding light-absorption data. Whenever possible, both types of spectra should be resolved into the component transitions, although the resolutions may not be unique. Moreover, complementary physical data, such as those obtained by measurement of fluorescence, various magnetic data, hydrodynamic properties, etc. should be recorded and evaluated for the same conjugated protein systems. Magnetic ORD and CD is a developing method and may be expected to add molecular information for many systems.

It is also evident from the present review that for most of the numerous systems described, only qualitative estimates of the origin of the observed optical activity can be given. Moreover, deductions made usually remain tentative and, in many cases, results cannot be interpreted unambiguously either by a single mechanism or by a superposition of more than one mechanism. Great caution must, therefore, be exercised in interpretation of experimental data, which frequently relies on chemical intuition. Undoubtedly, increased efforts will be made in the future to provide a more quantitative basis for results obtained, in particular for proteins whose three-dimensional structure will be determined with a sufficient degree of resolution.

Acknowledgements. This work was supported in part by a grant (No. 11 1772) from the Stiftung Volkswagenwerk. Thanks are due to Profs. *H. Sund* and *I. Z. Steinberg* for valuable comments.

VII. References

1. *Urnes, P., Doty, P.:* Advan. Protein Chem. *16*, 401 (1961).
2. *Schellman, J. A., Schellman, C.:* In: The proteins, Vol. II, p. 1 (ed. *H. Neurath*). New York: Academic Press 1964.
3. *Beychok, S.:* Science *154*, 1288 (1966).
4. *Fasman, G. D.* (ed.): Poly-α-amino acids. New York: Dekker 1967.
5. *Blout, E. R., Carver, J. P., Shechter, E.:* In: Optical rotatory dispersion and circular dichroism in organic chemistry, pp. 224 (ed. *G. Snatzke*). London: Heyden 1967.
6. *Eyring, H., Liu, H. C., Caldwell, D.:* Chem. Rev. *68*, 525 (1968).
7. *Moffitt, W.:* J. Chem. Phys. *25*, 467 (1956).
8. *Schellman, J. A.:* Accounts Chem. Res. *1*, 144 (1968).
9. *Deutsche, C. W., Lightner, D. A., Woody, R. W., Moscowitz, A.:* Ann. Rev. Phys. Chem. *20*, 407 (1969).
10. *Ulmer, D. D., Vallee, B. L.:* Advances in enzymology, Vol. 27, p. 37 (ed. *F. F. Nord*). New York: Interscience 1965.
11. *Gratzer, W. B., Cowburn, D. A.:* Nature *222*, 426 (1969).
12. *Urry, D. W.* (ed.): Spectroscopic approaches to biomolecular conformation. Chicago: American Medical Assoc. 1970.
13. *Perrin, J. H., Hart, P. A.:* J. Pharm. Sci. *59*, 431 (1970).
14. *Vallee, B. L., Wacker, W. E. C.:* In: The proteins, Vol. V (ed. *H. Neurath*). New York: Academic Press 1970.
15. *Beychok, S.:* Ann. Rev. Biochem. *37*, 437 (1968).
16. *Gratzer, W. B.:* In: Procedures in nucleic acid research, Vol. 2, p. 3 (eds. *G. L. Cantoni* and *D. R. Davies*). New York: Harper and Row 1971.
17. *Tinoco, I., Jr., Cantor, C. R.:* In: Methods of biochem. analysis, Vol. 18, p. 81 (ed. *D. Glick*). New York: Interscience 1970.
18. *Moscowitz, A.:* In: Optical rotatory dispersion, p. 150 (ed. *C. Djerassi*). New York: McGraw-Hill 1960.
19. — Rev. Mod. Phys. *32*, 440 (1960).
20. — Advan. Chem. Phys. *4*, 67 (1962).
21. *Tinoco, I., Jr.:* Advan. Chem. Phys. *4*, 113 (1962).
22. *Mason, S. F.:* Quart. Rev. *17*, 20 (1963).
23. *Brewster, J. H.:* In: Topics in stereochemistry, Vol. 2, p. 1 (eds. *N. L. Allinger* and *E. L. Eliel*). New York: Interscience 1967.
24. A Discussion on Circular Dichroism: Electronic and Structural Principles. Proc. Roy. Soc. (London) Ser. A *297*, pp. 2—172 (1967).
25. *Snatzke, G.:* Angew. Chem., Intern. Ed. Engl. 7, 14 (1968).

26. *Tinoco, I., Jr.*: In: Molecular biophysics, p. 269 (eds. *B. Pullman* and *M. Weissbluth*). New York–London: Academic Press 1965.
27. *Kirkwood, J. G.*: J. Chem. Phys. *5*, 479 (1937).
28. *Kasha, M.*: Radiation Res. *20*, 55 (1963).
29. *Kuhn, W.*: Trans. Faraday Soc. *26*, 293 (1930).
30. *Condon, E. V., Altar, W., Eyring, H.*: J. Chem. Phys. *5*, 753 (1937).
31. *Blout, E. R.*: Biopolym. Symp. *1*, 397 (1964).
32. *Djerassi, C.* (ed.): Optical rotatory dispersion: Applications to organic chemistry. New York: McGraw-Hill 1960.
33. *Klyne, W., Parker, A. C.*: In: Physical methods of organic chemistry, Vol. I, Part III, Edit. 3, p. 2335 (ed. *A. Weissberger*). New York: Interscience 1960.
34. *Heller, W., Fitts, D. D.*: In: Physical methods of organic chemistry, Vol. 1, Part III, p. 2147; Edit. 3 (ed. *A. Weissberger*). New York: Interscience 1960.
35. *Velluz, L., Legrand, M., Grosjean, M.*: Optical circular dichroism. Weinheim: Verlag Chemie. New York: Academic Press 1965.
36. *Crabbé, P.*: Optical rotatory dispersion and circular dichroism in organic chemistry. San Francisco: Holden-Day 1965.
37. *Beychok, S.*: In: Poly-α-amino acids, p. 293 (ed. *G. D. Fasman*). New York: Dekker 1967.
38. *Snatzke, G.* (ed.): Optical rotatory dispersion and circular dichroism in organic chemistry. London: Heyden 1967.
39. *Adkins, B. J., Yang, J. T.*: Biochemistry *7*, 266 (1968).
40. *Cassim, J. Y., Yang, J. T.*: Biochemistry *8*, 1947 (1969).
41. *Chau, K. H., Yang, J. T.*: Anal. Biochem. *46*, 616 (1972).
42. *Shashoua, V. E.*: J. Am. Chem. Soc. *86*, 2109 (1964).
43. — In: Hemes and hemoproteins, p. 93 (eds. *B. Chance, R. W. Estabrook* and *T. Yonetani*). New York–London: Academic Press 1966.
44. — Symp. Faraday Soc. *3*, 61 (1969).
45. *Foss, J. G., McCarville, M. E.*: J. Am. Chem. Soc. *87*, 228 (1965).
46. *Moscowitz, A.*: Proc. Roy. Soc. (London) Ser. A *297*, 16 (1967).
47. *Schatz, P. N., McCaffery, A. J.*: Quart. Rev. *23*, 552 (1969).
48. *Bayer, E., Bacher, A., Krauss, P., Voelter, W., Barth, G., Bunnenberg, E., Djerassi, C.*: Europ. J. Biochem. *22*, 580 (1971).
49. *Caldwell, D., Thorne, J. M., Eyring, H.*: Ann. Rev. Phys. Chem. *22*, 259 (1971).
50. *Timasheff, S. N., Gorbunoff, M. J.*: Ann. Rev. Biochem. *36*, 13 (1967).
51. *Yang, J. T.*: In: Poly-α-amino acids, p. 239 (ed. *G. D. Fasman*). New York: Dekker 1967.
52. *Jirgensons, B.*: Optical rotatory dispersion of proteins and other macromolecules. Berlin–Heidelberg–New York: Springer 1969.
53. *Moffitt, W., Yang, J. T.*: Proc. Natl. Acad. Sci. U. S. *42*, 596 (1956).
54. *Shechter, E., Blout, E. R.*: Proc. Natl. Acad. Sci. U. S. *51*, 695, 794 (1964).
55. *Carver, J. P., Shechter, E., Blout, E. R.*: J. Am. Chem. Soc. *88*, 2562 (1966).
56. *Yang, J. T.*: Proc. Natl. Acad. Sci. U. S. *53*, 438 (1965).
57. *Tinoco, I., Jr., Woody, R. W., Bradley, D. F.*: J. Chem. Phys. *38*, 1317 (1963).
58. *Vournakis, J. N., Yan, J. F., Scheraga, H. A.*: Biopolymers *6*, 1531 (1968).
59. *Tomimatsu, Y., Vitello, L., Gaffield, W.*: Biopolymers *4*, 653 (1966).
60. *Straus, J. H., Gordon, A. S., Wallach, D. F. H.*: Europ. J. Biochem. *11*, 201 (1969).
61. *Jirgensons, B.*: Biochim. Biophys. Acta *200*, 9 (1970).
62. *Javaherian, K., Beychok, S.*: J. Mol. Biol. *37*, 1 (1968).
63. *Fasman, G. D., Bodenheimer, E., Lindblow, C.*: Biochemistry *3*, 1665 (1964).
64. *Hashizume, H., Shiraki, M., Imahori, K.*: J. Biochem. (Tokyo) *62*, 543 (1967).

65. *Tiffany, M. L., Krimm, S.:* Biopolymers *8*, 347 (1969).
66. *Dearborn, D. G., Wetlaufer, D. B.:* Biochem. Biophys. Res. Commun. *39*, 314 (1970).
67. *Beychok, S.:* In: Probes of structure and function of macromolecules and membranes, Vol. 1, p. 143 (eds. *B. Chance, C. Lee* and *J. K. Blasie*). New York: Academic Press 1971.
68. *Chen, Y. H., Yang, J. T., Martinez, H .M.:* Biochemistry *11*, 4120 (1972).
69. *Saxena, V. P., Wetlaufer, D. B.:* Proc. Natl. Acad. Sci. U. S. *68*, 969 (1971).
70. *Greenfield, N., Fasman, G. D.:* Biochemistry *8*, 4108 (1969).
71. *Rosenkranz, H., Scholtan, W.:* Hoppe-Seylers Z. Physiol. Chem. *352*, 896 (1971).
72. *Grosse, R., Malur, J., Repke, K. R. H.:* FEBS Letters *25*, 313 (1972).
73. *Dalgleish, D. G.:* FEBS Letters *24*, 134 (1972).
74. *Hooker, T. M., Jr., Schellman, J. A.:* Biopolymers *9*, 1319 (1970)
75. *Blout, E. R., Stryer, L.:* Proc. Natl. Acad. Sci. U. S. *45*, 1591 (1959).
76. *Stryer, L., Blout, E. R.:* J. Am. Chem. Soc. *83*, 1411 (1961).
77. *Yamaoka, K., Resnik, R. A.:* Biopolymers *8*, 289 (1969).
78. *Greenblatt, C. L., Sharpless, N. E., Yamaoka, K.:* Mol. Pharmacol. *6*, 649 (1970).
79. *Myhr, B. C., Foss, J. G.:* Biopolymers *4*, 949 (1966).
80. *Eyring, E. J., Kraus, H., Yang, J. T.:* Biopolymers *6*, 703 (1968).
81. *Ballard, R. E., McCaffery, A. J., Mason, S. F.:* Biopolymers *4*, 97 (1966).
82. *Blauer, G.:* Nature *189*, 396 (1961).
83. — Biochim. Biophys. Acta *79*, 547 (1964).
84. *Stryer, L.:* Biochim. Biophys. Acta. *54*, 395 (1961).
85. *Blauer, G., Ehrenberg, A.:* Acta Chem. Scand *17*, 8 (1963).
86. — *Zvilichovsky, B.:* In: Structure and function of oxidation-reduction enzymes, p. 205 (eds. *Å. Åkeson* and *A. Ehrenberg*). Oxford: Pergamon 1972.
87. — *Alfassi, Z. B.:* Biochim. Biophys. Acta *133*, 206 (1967).
88. *Rifkind, J. M., Eichhorn, G. L.:* Biochemistry *9*, 1753 (1970).
89. *Wagner, K. G., Wulff, K.:* Biochem. Biophys. Res. Commun. *41*, 813 (1970).
90. *Cohen, P., Kidson, C.:* J. Mol. Biol. *35*, 241 (1968).
91. *Shapiro, J. T., Leng, M., Felsenfeld, G.:* Biochemistry *8*, 3219 (1969).
92. *Haynes, M., Garrett, R. A., Gratzer, W. B.:* Biochemistry *9*, 4410 (1970).
93. *De Santis, P., Savino, M.:* Biochim. Biophys. Acta *272*, 518 (1972).
94. *Rotilio, G., Agró, A., Calabrese, L., Bossa, F., Guerrieri, P., Mondovi, B.:* Biochemistry *10*, 616 (1971).
95. *Wood, E., Dalgleish, D., Bannister, W.:* Europ. J. Biochem. *18*, 187 (1971).
96. *Blumberg, W. E.:* In: The biochemistry of copper, p. 49 (eds. *J. Peisach, P. Aisen* and *W. E. Blumberg*). New York–London: Academic Press 1966.
97. *Hibino, Y., Samejima, T., Kajiyama, S., Nosoh, Y.:* Arch. Biochem. Biophys. *130*, 617 (1969).
98. *Falk, K. E., Reinhammar, B.:* Biochim. Biophys. Acta *285*, 84 (1972).
99. *Van Holde, K. E.:* Biochemistry *6*, 93 (1967).
100. *Takesada, H., Hamaguchi, K.:* J. Biochem. (Tokyo) *63*, 725 (1968).
101. *Nickerson, K. W., Van Holde, K. E.:* Comp. Biochem. Physiol. *39B*, 855 (1971).
102. *DePhillips, H. A., Nickerson, K. W., Johnson, M., VanHolde, K. E.:* Biochemistry *8*, 3665 (1969).
103. *DePhillips, H. A., Jr., Nickerson, K. W., Van Holde, K. E.:* J. Mol. Biol. *50*, 471 (1970).
104. *Ulmer, D. D., Vallee, B. L.:* Biochemistry *2*, 1335 (1963).
105. *Garbett, K., Darnall, D. W., Klotz, I. M., Williams, R. J. P.:* Arch. Biochem. Biophys. *103*, 419 (1969).

106. *Bossa, F., Brunori, M., Bates, G. W., Antonini, E., Fasella, P.:* Biochim. Biophys. Acta *207,* 41 (1970).
107. *Darnall, D. W., Garbett, K., Klotz, I. M., Aktipis, S., Keresztes-Nagy, S.:* Arch. Biochem. Biophys. *133,* 103 (1969).
108. *Blumberg, W. E., Peisach, J.:* Biochim. Biophys. Acta *126,* 269 (1966).
109. *Peisach, S., Levine, W. G., Blumberg, W. E.:* J. Biol. Chem. *242,* 2847 (1967).
110. *Morita, Y., Wadano, A., Ida, S.:* Agr. Biol. Chem. *35,* 255 (1971).
111. *Stigbrand, T., Sjöholm, I.:* Biochim. Biophys. Acta *263,* 244 (1972).
112. *Tang, S. P. W., Coleman, J. E., Myer, Y. P.:* J. Biol. Chem. *243,* 4286 (1968).
113. *Brill, A. S., Bryce, G. F., Maria, H. J.:* Biochim. Biophys. Acta *154,* 342 (1968).
114. *Nagy, B., Lehrer, S. S.:* Arch. Biochem. Biophys. *148,* 27 (1972).
115. *Tomimatsu, Y., Vickery, L. E.:* Biochim. Biophys. Acta *285,* 72 (1972).
116. *Gillard, R. D., McKenzie, E. D., Mason, R., Mayhew, S. G., Peel, J. L., Stangroom, J. E.:* Nature *208,* 769 (1965).
117. *Atherton, N. M., Garbett, K., Gillard, R. D., Mason, R., Mayhew, S. S., Peel, J. L., Stangroom, J. E.:* Nature *212,* 590 (1966).
118. *Kimura, T., Suzuki, K.:* J. Biol. Chem. *242,* 485 (1967).
118a. *Eaton, W. A., Lovenberg, W.:* J. Am. Chem. Soc. *92,* 7195 (1970).
119. *Palmer, G., Brintzinger, H., Estabrook, R. W.:* Biochemistry *6,* 1658 (1967).
120. *Garbett, K., Gillard, R. D., Knowles, P. F., Stangroom, J. E.:* Nature *215,* 824 (1967).
121. *Marlborough, D. I., Hall, D. O., Cammack, R.:* Biochem. Biophys. Res. Commun. *35,* 410 (1969).
122. *Sutherland, J. C., Salmeen, I., Sun, A. S. K., Klein, M. P.:* Biochim. Biophys. Acta *263,* 550 (1972).
123. *Palmer, G., Massey, V.:* J. Biol. Chem. *244,* 2614 (1969).
124. *Multani, J. S., Mortenson, L. E.:* Biochim. Biophys. Acta *256,* 66 (1972).
125. *Listowsky, I., Blauer, G., Englard, S., Betheil, J. J.:* Biochemistry *11,* 2176 (1972).
126. *Pulido, P., Kägi, J. H. R., Vallee, B. L.:* Biochemistry *5,* 1768 (1966).
127. *Behnke, W. D., Vallee, B. L.:* Proc. Natl. Acad. Sci. U. S. *69,* 2442 (1972).
128. *Coleman, J. E.:* J. Biol. Chem. *243,* 4574 (1968).
129. *Einarsson, R., Zeppezauer, M.:* Acta Chem. Scand. *24,* 1098 (1970).
130. *Henkens, R. W., Sturtevant, J. M.:* Biochemistry *11,* 206 (1972).
131. *Coleman, J. E., Coleman, R. V.:* J. Biol. Chem. *247,* 4718 (1972).
132. *Perutz, M. F., Muirhead, M., Cox, J. M., Goaman, L. C. G.:* Nature *219,* 131 (1968).
133. *Kendrew, J. C., Watson, H. C., Strandberg, B. E., Dickerson, R. E., Phillips, D. C., Shore, V. C.:* Nature *190,* 666 (1961).
134. *Antonini, E., Brunori, M.:* Ann. Rev. Biochem. *39,* 977 (1970).
135. *Dickerson, R. E.:* Ann. Rev. Biochem. *41,* 815 (1972).
136. *Hsu, M. C., Woody, R. W.:* J. Am. Chem. Soc. *91,* 3679 (1969).
137. — — J. Am. Chem. Soc. *93,* 3515 (1971).
138. *Beychok, S., Blout, E. R.:* J. Mol. Biol. *3,* 769 (1961).
139. *Eichhorn, G. L.:* Tetrahedron *13,* 208 (1961).
140. *Samejima, T., Yang, J. T.:* J. Mol. Biol. *8,* 863 (1964).
141. *Willick, G. E., Schonbaum, G. R., Kay, C. M.:* Biochemistry *8,* 3729 (1969).
142. *Nagai, M., Sugita, Y., Yoneyama, Y.:* J. Biol. Chem. *244,* 1651 (1969).
143. *Ruckpaul, K., Rein, H., Ristau, O., Jung, F.:* Biochim. Biophys. Acta *221,* 9 (1970).
144. *Sugita, Y., Nagai, M., Yoneyama, Y.:* J. Biol. Chem. *246,* 383 (1971).

145. *Geraci, G., Li, T. K.:* Biochemistry *8*, 1848 (1969).
146. *Goodall, P. T., Shooter, E. M.:* J. Mol. Biol. *39*, 675 (1969).
147. *Ueda, Y., Shiga, T., Tyuma, I.:* Biochem. Biophys. Res. Commun. *35*, 1 (1969).
148. — — — Biochim. Biophys. Acta *207*, 18 (1970).
149. *Beychok, S.:* Biopolymers *2*, 575 (1964).
150. *Melki, G.:* J. Phys. (Paris) *33*, 37 (1972).
151. *Lampe, J., Rein, H., Scheler, W.:* FEBS Letters *23*, 282 (1972).
152. *Ruckpaul, K., Rein, H., Jung, F.:* Acta Biol. Med. Ger. *24*, 445 (1970).
153. *Formanek, H., Engel, J.:* Biochim. Biophys. Acta *160*, 151 (1968).
154. *Wollmer, A., Sick, H., Gersonde, K.:* Europ. J. Biochem. *16*, 303 (1970).
155. *Gersonde, K., Wollmer, A.:* Europ. J. Biochem. *15*, 226 (1970).
156. *Wollmer, A., Buse, G., Sick, H., Gersonde, K.:* Europ. J. Biochem. *24*, 547 (1972).
157. *Li, T. K., Johnson, B. P.:* Biochemistry *8*, 3638 (1969).
158. *Bolard, J., Garnier, A.:* Biochim. Biophys. Acta *263*, 535 (1972).
159. *Volkenstein, M. V., Sharonov, J. A., Shemelin, A. K.:* Nature *209*, 709 (1966).
160. *Brunori, M., Giacometti, G. M., Antonini, E., Wyman, J.:* J. Mol. Biol. *63*, 139 (1972).
161. *Blauer, G., Harmatz, D., Zvilichovsky, B.:* FEBS Letters, in press.
162. *Beychok, S., Tyuma, I., Benesch, R. E., Benesch, R.:* J. Biol. Chem. *242*, 2460 (1967).
163. *Urry, D. W.:* J. Biol. Chem. *242*, 4441 (1967).
164. *Simon, S. R., Cantor, C. R.:* Proc. Natl. Acad. Sci. U. S. *63*, 205 (1969).
165. *Ruckpaul, K., Rein, H., Ristau, O., Jung, F.:* Experientia *26*, 1079 (1970).
166. *Urry, D. W.:* In: Structure and function of cytochromes, p. 311 (eds. *K. Okunuki, M. D. Kamen* and *I. Sekuzu*). Tokyo: University of Tokyo Press 1968.
167. *Breslow, E., Beychok, S., Hardman, K. D., Gurd, F. R. N.:* J. Biol. Chem. *240*, 304 (1965).
168. *Harrison, S. C., Blout, E. R.:* J. Biol. Chem. *240*, 299 (1965).
169. *Urnes, P. J., Imahori, K., Doty, P.:* Proc. Natl. Acad. Sci. U. S. *47*, 1635 (1961).
170. *Samejima, T., Kita, M.:* J. Biochem. (Tokyo) *65*, 759 (1969).
171. *Kendrew, J. C.:* Science *139*, 1259 (1963).
172. *Ruckpaul, K., Grill, H., Jung, F.:* Acta Biol. Med. Ger. *24*, 25 (1970).
173. *Asakura, T., Minakami, S., Yoneyama, Y., Yoshikawa, H.:* J. Biochem. (Tokyo) *56*, 594 (1964).
174. *Brunori, M., Engel, J., Schuster, T. M.:* J. Biol. Chem. *242*, 773 (1967).
175. *Urry, D. W.:* Proc. Natl. Acad. Sci. U. S. *54*, 640 (1965).
176. *Schellman, J. A., King, T. E.:* In: Hemes and hemoproteins, p. 507 (eds. *B. Chance, R. W. Estabrook* and *T. Yonetani*). New York–London: Academic Press (1966).
177. *Hänisch, G., Engel, J., Brunori, M., Fasold, H.:* Europ. J. Biochem. *9*, 335 (1969).
178. *Li, T. K., Johnson, B. P.:* Biochemistry *8*, 2083 (1969).
179. *Sugita, Y., Dohi, Y., Yoneyama, Y.:* Biochem. Biophys. Res. Commun. *31*, 447 (1968).
180. *Devi, S. K.:* In: Genetical, functional and physical studies of hemoglobin, p. 262 (eds. *T. Arends, G. Bemski* and *R. L. Nagel*). Basel: Karger 1971.
181. *Melki, G.:* Biochim. Biophys. Acta *263*, 226 (1972).
182. *Nagai, M., Sugita, Y., Yoneyama, Y.:* J. Biol. Chem. *247*, 285 (1972).
183. *Murayama, M.:* Nature *194*, 933 (1962).
184. *Sasazuki, T., Isomoto, A., Nakajima, H.:* J. Mol. Biol. *65*, 365 (1972).

185. *Hamaguchi, H., Isomoto, A., Nakajima, H.:* Biochem. Biophys. Res. Commun. *35,* 6 (1969).
186. *— — Miyake, Y., Nakajima, H.:* Biochemistry *10,* 1741 (1971).
187. *Makinen, M. W., Kon, H.:* Biochemistry *10,* 43 (1971).
188. *Sasazuki, T., Isomoto, A.:* Proc. Japan Acad. *46,* 1060 (1970).
189. *Waks, M., Kahn, P. C., Beychok, S.:* Biochem. Biophys. Res. Commun. *45,* 1232 (1971).
190. *Ruckpaul, K., Rein, H., Jung, F.:* Naturwissenschaften *57,* 131 (1970).
191. *Breslow, E., Koehler, R., Girotti, A. W.:* J. Biol. Chem. *242,* 4149 (1967).
192. *Fabry, T. L., Simo, C., Javaherian, K.:* Biochim. Biophys. Acta *160,* 118 (1968).
193. *Andres, S. F., Atassi, M. Z.:* Biochemistry *9,* 2268 (1970).
194. *Hsu, M. C., Woody, R. W.:* Biopolymers *10,* 1421 (1971).
195. *Stryer, L.:* J. Mol. Biol. *13,* 482 (1965).
196. *Urry, D. W., Doty, P.:* J. Am. Chem. Soc. *87,* 2756 (1965).
197. *Ulmer, D. D.:* Biochemistry *5,* 1886 (1966).
198. *Mirsky, R., George, P.:* Proc. Natl. Acad. Sci. U. S. *56,* 222 (1966).
199. *Ulmer, D. D.:* Biochemistry *4,* 902 (1965).
200. *Vinogradov, S., Zand, R.:* Arch. Biochem. Biophys. *125,* 902 (1968).
201. *Myer, Y. P.:* J. Biol. Chem. *243,* 2115 (1968).
202. *Zand, R., Vinogradov, S.:* Biochem. Biophys. Res. Commun. *26,* 121 (1967).
203. *Myer, Y. P.:* Biochim. Biophys. Acta *154,* 84 (1968).
204. *Zand, R., Vinogradov, S.:* Arch. Biochem. Biophys. *125,* 94 (1968).
205. *Dickerson, R. E., Takano, T., Eisenberg, D., Kallai, O. B., Samson, L., Cooper, A., Margoliash, E.:* J. Biol. Chem. *246,* 1511 (1971).
206. *Eaton, W. A., Charney, E.:* J. Chem. Phys. *51,* 4502 (1969).
207. *Myer, Y. P., Harbury, H. A.:* Proc. Natl. Acad. Sci. U. S. *54,* 1391 (1965).
208. *Harbury, H. A., Myer, Y. P., Murphy, A. J., Vinogradov, S. N.:* In: Hemes and hemoproteins, p. 415 (eds. *B. Chance, R. W. Estabrook* and *T. Yonetani*). New York: Academic Press 1966.
209. *Myer, Y. P., Murphy, A. J., Harbury, H. A.:* J. Biol. Chem. *241,* 5370 (1966).
210. *Greenwood, C.:* In: Structure and function of cytochromes, p. 340 (eds. *K. Okunuki, M. D. Kamen* and *I. Sekuzu*). Tokyo: University of Tokyo Press 1968.
211. *Myer, Y. P.:* Biochim. Biophys. Acta *214,* 94 (1970).
212. *Mirsky, R., George, P.:* Biochemistry *6,* 3671 (1967).
213. *Hamaguchi, K., Ikeda, K., Narita, K.:* In: Structure and function of cytochromes, p. 328 (eds. *K. Okunuki, M. D. Kamen* and *I. Sekuzu*). Tokyo: University of Tokyo Press 1968.
214. *Van Gelder, B. F., Urry, D. W., Beinert, H.:* In: Structure and function of cytochromes, p. 335 (eds. *K. Okunuki, M. D. Kamen* and *I. Sekuzu*). Tokyo: University of Tokyo Press 1968.
215. *Imai, Y., Imai, K., Ikeda, K., Hamaguchi, K., Horio, T.:* J. Biochem. (Tokyo) *65,* 629 (1969).
216. *Flatmark, T., Robinson, A. B.:* In: Structure and function of cytochromes, p. 318 (eds. *K. Okunuki, M. D. Kamen* and *I. Sekuzu*). Tokyo: University of Tokyo Press 1968.
217. *Yong, F. C., King, T. E.:* J. Biol. Chem. *245,* 1331 (1970).
218. *Drucker, H., Campbell, L. L., Woody, R. W.:* Biochemistry *9,* 1519 (1970).
219. *Flatmark, T.:* J. Biol. Chem. *242,* 2454 (1967).
220. *Ulmer, D. D., Vallee, B. L., Gorchein, A., Neuberger, A.:* Nature *206,* 825 (1965).

221. *Myer, Y. P.:* Biochemistry 7, 765 (1968).
222. *Kaminsky, L. S., Yong, F. C., King, T. E.:* J. Biol. Chem. *247*, 1354 (1972).
223. *Myer, Y. P.:* Biochemistry *11*, 4203 (1972).
224. *Ulmer, D. D.:* Proc. Natl. Acad. Sci. U. S. *55*, 894 (1966).
225. *Saludjian, P., Shechter, E.:* Biopolymers *5*, 561 (1967).
226. *Myer, Y. P., Harbury, H. A.:* J. Biol. Chem. *241*, 4299 (1966).
227. *Urry, D. W.:* J. Am. Chem. Soc. *89*, 4190 (1967).
228. — *Pettegrew, J. W.:* J. Am. Chem. Soc. *89*, 5276 (1967).
229. *Shashoua, V. E.:* Nature *203*, 972 (1964).
230. — Arch. Biochem. Biophys. *111*, 550 (1965).
231. *Greenwood, C., Wilson, M. T.:* Europ. J. Biochem. *22*, 5 (1971).
232. *Shooley, D. A., Bunnenberg, E., Djerassi, C.:* Proc. Natl. Acad. Sci. U. S. *53*, 579 (1965).
233. *Sutherland, J. C., Klein, M. P.:* J. Chem. Phys. *57*, 76 (1972).
234. *Morrison, M., Duffield, J.:* In: Hemes and hemoproteins, p. 431 (eds. *B. Chance*, *R. W. Estabrook* and *T. Yonetani*). New York: Academic Press 1966.
235. *Shashoua, V. E., Estabrook, R. W.:* In: Hemes and hemoproteins, p. 427 (eds. *B. Chance, R. W. Estabrook* and *T. Yonetani*). New York: Academic Press 1966.
236. *Yong, F. C., King, T. E., Oldham, S., Waterman, M. R., Mason, H. S.:* Arch. Biochem. Biophys. *138*, 96 (1970).
237. *Schnellbacher, E., Lumper, L.:* Hoppe-Seylers Z. Physiol. Chem. *352*, 615 (1971).
238. *Huntley, T. E., Strittmatter, P.:* J. Biol. Chem. *247*, 4641 (1972).
239. — — J. Biol. Chem. *247*, 4648 (1972).
240. *Okada, Y., Okunuki, K.:* J. Biochem. (Tokyo) *67*, 487 (1970).
241. — — J. Biochem. (Tokyo) *67*, 603 (1970).
242. *Chance, B., Estabrook, R. W., Yonetani, T.:* (eds.) Hemes and hemoproteins. New York: Academic Press 1966.
243. *Urry, D. W., Wainio, W. W., Grebner, D.:* Biochem. Biophys. Res. Commun, *27*, 625 (1967).
244. — *Van Gelder, B. F.:* In: Structure and function of cytochromes, p. 210 (eds. *K. Okunuki, M. D. Kamen* and *I. Sekuzu*). Tokyo: University of Tokyo Press 1968.
245. *King, T. E., Yong, F. C., Bayley, P. M.:* In: Structure and function of cytochromes, p. 204 (eds. *K. Okunuki, M. D. Kamen* and *I. Sekuzu*). Tokyo: University of Tokyo Press 1968.
246. *Myer, Y. P., King, T. E.:* Biochem. Biophys. Res. Commun. *33*, 43 (1968).
247. *King, T. E., Bayley, P. M., Yong, F. C.:* Europ. J. Biochem. *20*, 103 (1971).
248. *Yong, F. C., King, T. E.:* Biochem. Biophys. Res. Commun. *40*, 1445 (1970).
249. *Morrison, M.:* In: Structure and function of cytochromes, p. 89 (eds. *K. Okunuki, M. D. Kamen* and *I. Sekuzu*). Tokyo: University of Tokyo Press 1968.
250. *Myer, Y. P., King, T. E.:* Biochem. Biophys. Res. Commun. *34*, 170 (1969).
251. *Love, B., Auer, H. E.:* Biochem. Biophys. Res. Commun. *41*, 1437 (1970).
252. *Yong, F. C., King, T. E.:* Europ. J. Biochem. *20*, 111 (1971).
253. — — J. Biol. Chem. *244*, 509, 515 (1969).
254. — *Bayley, P. M., King, T. E.:* In: Structure and function of cytochromes, p. 196 (eds. *K. Okunuki, M. D. Kamen* and *I. Sekuzu*). Tokyo: University of Tokyo Press 1968.
255. *Yang, J. T., Samejima, T.:* J. Biol. Chem. *238*, 3262 (1963).
256. *Osbahr, A. J., Eichhorn, G. L.:* J. Biol. Chem. *237*, 1820 (1962).
257. *Samejima, T., Kita, M.:* Biochim. Biophys. Acta *175*, 24 (1969).

258. *Strickland, E. H.:* Biochim. Biophys. Acta *151*, 70 (1968).
259. — *Kay, E., Shannon, L. M., Horwitz, J.:* J. Biol. Chem. *243*, 3560 (1968).
260. *Ellis, W. D., Dunford, H. B.:* Can. J. Biochem. *46*, 1231 (1968).
261. *Hamaguchi, K., Ikeda, K., Yoshida, C., Morita, Y.:* J. Biochem. (Tokyo) *66*, 191 (1969).
262. *Anderson, G. R., Jenner, E. L., Mumford, F. E.:* Biochim. Biophys. Acta *221*, 69 (1970).
263. *Kroes, H. H.:* Biochem. Biophys. Res. Commun. *31*, 877 (1968).
264. *Burke, M. J., Pratt, D. C., Moscowitz, A.:* Biochemistry *11*, 4025 (1972)
265. *Boucher, L. J., Crespi, H. L., Katz, J. J.:* Biochemistry *5*, 3796 (1966).
266. *Barrett, J.:* In: Structure and function of cytochromes, p. 701 (eds. *K. Okunuki, M. D. Kamen* and *I. Sekuzu*). Tokyo: University of Tokyo Press 1968.
267. *Moscowitz, A., Krueger, W. C., Kay, I. T., Skewes, G., Bruckenstein, S.:* Proc. Natl. Acad. Sci. U. S. *52*, 1190 (1964).
268. *Fujimori, E., Pecci, J.:* Biochim. Biophys. Acta *221*, 132 (1970).
269. *Philipson, K. D., Sauer, K.:* Biochemistry *11*, 1880 (1972).
270. *Sturtevant, J. M., Tsong, T. Y.:* J. Biol. Chem. *243*, 2359 (1968); *244*, 4942 (1969).
271. *Iwatsubo, M., Di Franco, A.:* In: Structure and function of cytochromes, p. 613 (eds. *K. Okunuki, M. D. Kamen* and *I. Sekuzu*). Tokyo: University of Tokyo Press 1968.
272. *Tsong, T. Y., Sturtevant, J. M.:* J. Biol. Chem. *244*, 2397 (1969).
272a. — — J. Am. Chem. Soc *91*, 2382 (1969).
273. *Iwatsubo, M., Risler, J. L.:* Europ. J. Biochem. *9*, 280 (1969).
274. *Briat, B., Berger, D., Leliboux, M.:* J. Chem. Phys. *57*, 5606 (1972).
275. *Tollin, G.:* Biochemistry *7*, 1720 (1968).
276. *Edmonson, D. E., Tollin, G.:* Biochemistry *10*, 113 (1971).
277. — — Biochemistry *10*, 124 (1971).
278. *D'Anna, J. A., Tollin, G.:* Biochemistry *11*, 1073 (1972).
279. *Brady, A. H., Beychok, S.:* J. Biol. Chem. *244*, 4634 (1969).
280. — — J. Biol. Chem. *246*, 5498 (1971).
281. *Aki, K., Takagi, T., Isemura, T., Yamano, T.:* Biochim. Biophys. Acta *122*, 193 (1966).
282. *Takagi, T., Aki, K., Isemura, T., Yamano, T.:* Biochem. Biophys. Res. Commun. *24*, 501 (1966).
283. *Kotaki, A., Sugiura, N., Yagi, K.:* Biochim. Biophys. Acta *151*, 689 (1968).
284. *Wellner, D.:* Biochemistry *5*, 1585 (1966).
285. *Aleman, V., Handler, P.:* J. Biol. Chem. *242*, 4087 (1967).
286. *Listowsky, I., Furfine, C. S., Betheil, J. J., Englard, S.:* J. Biol. Chem. *240*, 4253 (1965).
287. *De Vijlder, J. J. M., Boers, W., Hilvers, A. G., Harmsen, B. J. M., Slater, E. C.:* In: Pyridine nucleotide-dependent dehydrogenases, p. 233 (ed. *H. Sund*). Berlin–Heidelberg–New York: Springer 1970.
288. — *Harmsen, B. J. M.:* Biochim. Biophys. Acta *178*, 434 (1969).
289. *Bayley, P. M., Radda, G. K.:* Biochem. J. *94*, 31 P (1965).
290. *Magar, M. E.:* Biochim. Biophys. Acta *96*, 345 (1965).
291. *Jaenicke, R., Gratzer, W. B.:* Europ. J. Biochem. *10*, 158 (1969).
292. — In: Pyridine nucleotide-dependent dehydrogenases, p. 209 (ed. *H. Sund*). Berlin–Heidelberg–New York: Springer 1970.
293. *Heck, H. d'A.:* J. Biol. Chem. *244*, 4375 (1969).
294. *Koberstein, R., Sund, H.:* FEBS Letters *19*, 149 (1971).

295. *Rosenberg, A., Theorell, H., Yonetani, T.:* Arch. Biochem. Biophys. *110*, 413 (1965).
296. *Di Franco, A., Iwatsubo, M.:* Europ. J. Biochem. *30*, 517 (1972).
297. *Jallon, J. M., Iwatsubo, M.:* Biochem. Biophys. Res. Commun. *45*, 964 (1971).
298. *Sund, H., Koberstein, R., Krause, J., Markau, K.:* In: Proc. 1st. Eur. Biophys. Congr., Baden, Vol. VI, p. 97 (eds. *E. Broda, A. Locker* and *H. Springer-Lederer*). Vienna: Wiener Med. Akademie 1971.
299. *Drum, D. E., Vallee, B. L.:* Biochem. Biophys. Res. Commun. *41*, 33 (1970).
300. *Anderson, S. R.*, Biochemistry *10*, 4162 (1971).
301. *D'Souza, L., Freisheim, J. H.:* Biochemistry *11*, 3770 (1972).
302. *Miles, D. W., Urry, D. W.:* J. Biol. Chem. *243*, 4181 (1968).
303. *Secrist, J. A. III, Leonard, N. J.:* J. Am. Chem. Soc. *94*, 1702 (1972).
304. *Dowben, R. M., Orkin, S. H.:* Proc. Natl. Acad. Sci. U. S. *58*, 2051 (1967).
305. *Shechter, E.:* Europ. J. Biochem. *10*, 274 (1969).
306. *Blauer, G., King, T. E.:* Biochem. Biophys. Res. Commun. *31*, 678 (1968).
307. — — J. Biol. Chem. *245*, 372 (1970).
308. — *Harmatz, D., Snir, J.:* Biochim. Biophys. Acta *278*, 68 (1972).
309. — — — In: Proc. 1st Eur. Biophys. Congr., Baden, Vol. I, p. 127 (eds. *E. Broda, A. Locker* and *H. Springer-Lederer*). Vienna: Wiener Med. Akademie 1971.
310. — — Biochim. Biophys. Acta *278*, 89 (1972).
311. *Woolley, P. V. III., Hunter, M. J.:* Arch. Biochem. Biophys. *140*, 197 (1970).
312. *Blauer, G., Harmatz, D., Naparstek, A.:* FEBS Letters *9*, 53 (1970).
313. *Mason, S. F.:* In: Optical rotatory dispersion and circular dichroism in organic chemistry, p. 71 (ed. *G. Snatzke*). London: Heyden 1967.
314. *Blauer, G., Harmatz, D.:* In: Jerusalem Symp. Quantum Chem. Biochem., Vol. V, p. 709 (eds. *E. Bergmann* and *B. Pullman*). Jerusalem: Isr. Acad. Sci. Human. 1973.
315. — *Blondheim, S. H., Harmatz, D., Kapitulnik, J., Kaufmann, N. A., Zvilichovsky, B.:* FEBS Letters, in press.
316. *Perrin, J. H., Wilsey, M.:* J. Chem. Soc., *D*, 769 (1971).
317. *Blauer, G., Zvilichovsky, B.:* Israel J. Chem. (1973), in press.
318. *Brodersen, R., Funding, L., Pedersen, A. O., Röigaard-Petersen, H.:* Scand. J. Clin. Lab. Invest. *29*, 433 (1972).
319. *Chignell, C. F.:* Mol. Pharm. *5*, 244 (1969).
320. — Mol. Pharm. *6*, 1 (1970).
321. *Perrin, J. H., Nelson, D. A.:* Life Sci. *11*, 277 (1972).
322. *Chignell, C. F.:* Mol. Pharm. *5*, 455 (1969).
323. *Sjöholm, I., Grahnén, A.:* FEBS Letters *22*, 109 (1972).
324. *Anderson, S. R.:* Biochemistry *8*, 4838 (1969).
325. *Bradshaw, R. A., Shearer, W. T., Gurd, F. R. N.:* J. Biol. Chem. *243*, 3817 (1968).
326. *Tsangaris, J. M., Chang, J. W., Martin, R. B.:* Arch. Biochem. Biophys. *130*, 53 (1969).
327. *Shaw, T. I.:* In: Handbook of sensory physiology, Vol. 7 (1), p. 180 (ed. *H. J. A. Dartnall*). Berlin–Heidelberg–New York: Springer 1972.
328. *Crescitelli, F., Mommaerts, W. F. H. M., Shaw, T. I.:* Proc. Natl. Acad. Sci. U. S. *56*, 1729 (1966).
329. *Williams, T. P.:* Vision Res. *6*, 293 (1966).
330. *Shichi, H., Lewis, M. S., Irreverre, F., Stone, A. L.:* J. Biol. Chem. *244*, 529 (1969).
331. *Hubbard, R., Bownds, D., Yoshizawa, T.:* Cold Spring Harbor Symp. Quant. Biol. *30*, 301 (1965).

332. *Kito, Y., Takezaki, M.:* Nature *211*, 197 (1966).
333. *Takezaki, M., Kito, Y.:* Nature *215*, 1197 (1967).
334. *Crescitelli, F., Foster, R. F., Shaw, T. I.:* J. Physiol. *202*, 189 (1969).
335. *Waggoner, A. S., Stryer, L.:* Biochemistry *10*, 3250 (1971).
336. *Kito, Y., Azuma, M., Maeda, Y.:* Biochim. Biophys. Acta *154*, 352 (1968).
337. *Johnston, E. M., Zand, R.:* Biochem. Biophys. Res. Commun. *47*, 712 (1972).
338. *Shichi, H.:* Biochemistry *9*, 1973 (1970).
339. *Horwitz, J., Heller, J.:* Biochemistry *10*, 1402 (1971).
340. *Shichi, H.:* Photochem. Photobiol. *13*, 499 (1971).
341. *Breslow, E.:* Proc. Natl. Acad. Sci. U. S. *67*, 493 (1970).
342. — *Aanning, H. L., Abrash, L., Schmir, M.:* J. Biol. Chem. *246*, 5179 (1971).
343. — *Weis, J.:* Biochemistry *11*, 3474 (1972).
344. *Pritchard, A. B., McCormic, D. B., Wright, L. D.:* Biochem. Biophys. Res. Commun. *25*, *524* (1966).
345. *Rockey, J. H., Dorrington, K. J., Montgomery, P. C.:* J. Immunol. Meth. *1*, 67 (1971).
346. *Cathou, R. E., Kulczycki, A., Jr., Haber, E.:* Biochemistry *7*, 3958 (1968).
347. *Holowka, D. A., Strosberg, A. D., Kimball, J. W., Haber, E., Cathou, R. E.:* Proc. Natl. Acad. Sci. U. S. *69*, 3399 (1972).
348. *Rockey, J. M., Dorrington, K. J., Montgomery, P. C.:* Nature *232*, 192 (1971).
349. *Scanu, A., Hirz, R.:* Proc. Natl. Acad. Sci. U. S. *59*, 890 (1968).
350. — — Nature *218*, 200 (1968).
351. *Gotto, A. M., Levy, R. I., Fredrickson, D. S.:* Proc. Natl. Acad. Sci. U. S. *60*, 1436 (1968).
352. *Murphy, A. J.:* Biochemistry *10*, 3723 (1971).
353. *Nagy, B., Strzelecka-Golaszewska, H.:* Arch. Biochem. Biophys. *150*, 428 (1972).
354. *Torchinsky, Y. M., Livanova, N. B., Pikhelgas, V. Y.:* Biochim. Biophys. Acta *110*, 619 (1965).
355. *Johnson, G. F., Graves, D. J.:* Biochemistry *5*, 2906 (1966).
356. *Adachi, O., Yamada, H.:* J. Biochem. (Tokyo) *65*, 639 (1969).
357. *Fasella, P., Hammes, G. G.:* Biochemistry *4*, 801 (1965).
358. *Breusov, Y. N., Ivanov, V. I., Karpeisky, M. Y., Morozov, Y. V.:* Biochim. Biophys. Acta *92*, 388 (1964).
359. *Huntley, T. E., Metzler, D. E.:* Biochem. Biophys. Res. Commun. *26*, 109 (1967).
360. *Kägi, J. H. R., Li, T.-K., Vallee, B. L.:* Biochemistry *10*, 1007 (1971).
361. *Ullrich, J., Wollmer, A.:* Hoppe-Seylers Z. Physiol. Chem. *351*, 132 (1970).
362. *Griffin, J. H., Rosenbusch, J. P., Weber, K. K., Blout, E. R.:* J. Biol. Chem. *247*, 6482 (1972).
363. *Glazer, A. N., Simmons, N. S.:* J. Am. Chem. Soc. *88*, 2335 (1966).
364. *Ikeda, K., Hamaguchi, K.:* J. Biochem. (Tokyo) *66*, 513 (1969).
365. *Cathou, R. E., Hammes, G. G., Schimmel, P. R.:* Biochemistry *4*, 2687 (1965).
366. *Samejima, T., Kita, M., Saneyoshi, M., Sawada, F.:* Biochim. Biophys. Acta *179*, 1 (1969).
367. *Yoshida, N., Kuriyama, K., Iwata, T., Otsuka, H.:* Biochem. Biophys. Res. Commun. *43*, 954 (1971).
368. *Oshima, T., Imahori, K.:* J. Biochem. (Tokyo) *70*, 197 (1971).
369. *Sander, C., Ts'o, P. O. P.:* Biochemistry *10*, 1953 (1971).
370. *Fairclough, G. F., Jr., Vallee, B. L.:* Biochemistry *9*, 4087 (1970).
371. — — Biochemistry *10*, 2470 (1971).
372. *Ohta, T., Shimada, I., Imahori, K.:* J. Mol. Biol. *26*, 519 (1967).

373. *Shih, T. Y., Fasman, G. D.:* J. Mol. Biol. *52,* 125 (1970).
374. *Tuan, D. Y. H., Bonner, J.:* J. Mol. Biol. *45,* 59 (1969).
375. *Šponar, J., Boublik, M., Frič, I., Šormová, Z.:* Biochim. Biophys. Acta *209,* 532 (1970).
376. *Frič, I., Šponar, J.:* Biopolymers *10,* 1525 (1971).
377. *Simpson, R. T., Sober, H. A.:* Biochemistry *9,* 3103 (1970).
378. *Henson, P., Walker, I. O.:* Europ. J. Biochem. *16,* 524 (1970).
379. *Matsuyama, A., Tagashira, Y., Nagata, C.:* Biochim. Biophys. Acta *240,* 184 (1971).
380. *Williams, R. E., Lurquin, P. F., Seligy, V. L.:* Europ. J. Biochem. *29,* 426 (1972).
381. *Li, H. J., Isenberg, I., Johnson, W. C., Jr.:* Biochemistry *10,* 2587 (1971).
382. *Wagner, T., Spelsberg, T. C.:* Biochemistry *10,* 2599 (1971).
383. *Wagner, T. E., Vandegrift, V.:* Biochemistry *11,* 1431 (1972).
384. *Slayter, H. S., Shih, T. Y., Adler, A. J., Fasman, G. D.:* Biochemistry *11,* 3044 (1972).
385. *Permogorov, V. I., Debabov, U. G., Sladkova, I. A., Rebentish, B. A.:* Biochim. Biophys. Acta *199,* 556 (1970).
386. *Ramm, E. I., Vorob'ev, V. I., Birshtein, T. M., Bolotina, I. A., Volkensthein, M. V.:* Europ. J. Biochem. *25,* 245 (1972).
387. *Shih, T. Y., Fasman, G. D.:* Biochemistry *11,* 398 (1972).
388. *Fasman, G. D., Schaffhausen, B., Goldsmith, L., Adler, A.:* Biochemistry *9,* 2814 (1970).
389. *Adler, A. J., Fasman, G. D.:* J. Phys. Chem. *75,* 1516 (1971).
390. *Olins, D. E., Olins, A. L.:* J. Mol. Biol. *57,* 437 (1971).
391. *Boffa, L., Saccomani, G., Tamburro, A. M., Scatturin, A., Vidali, G.:* Intern. J. Protein Res. *3,* 357 (1971).
392. *Bush, C. A., Scheraga, H. A.:* Biochemistry *6,* 3036 (1967).
393. *Srinivasan, S., Lauffer, M. A.:* Arch. Biochem. Biophys. *148,* 568 (1972).

Received May 19, 1973

Some Aspects of the Heteropolymolybdates and Heteropolytungstates

T. J. R. Weakley

Chemistry Department, Dundee University, Dundee DD1 4HN, Scotland

Table of Contents

1. Introduction

The iso- and hetero-polyanions of the vanadium and chromium subgroups have been the part-subjects of several monographs in recent years (*1, 2, 3*). Good accounts have been given of their structural chemistry (*4*), of the isopoly-vanadates, -niobates, and -tantalates (*5*), and of the isopolyanions (*6*) and heteropolyanions (*7*) of molybdenum and tungsten, though unfortunately the last two reviews may not be readily accessible to many workers. Here, we survey some aspects of the heteropoly-molybdates and -tungstates, particularly those which may interest workers in the fields of structural chemistry and the electronic properties of metal complexes. Kinetics, equilibria, and physico-chemical properties in solution, which we hope to treat separately, are excluded. In the hope of stimulating further work, we shall indicate which heteropolyanions are currently of unknown structure (and, in many cases, of uncertain formula). We have tried to cover the available literature up to July 1973, but for reasons of space the bibliography, particularly as regards the older work, is incomplete; many further references will be found in the papers cited here.

This review, then, deals with oxoanions containing atoms of two elements other than hydrogen[1]) in positive oxidation states; up to eighteen atoms of molybdenum (VI) or tungsten (VI), and one or two *central atoms* or *heteroatoms*. The heteroelement may be a non-metal of moderate electronegativity, or a metal (especially a transition metal). The only periodic groups *not* represented among the reported heteroelements are IA, IIA (other than beryllium), the halogens (other than iodine), and the rare gas elements. The classes of heteropolyanions of known structure are listed in Table 1, together with the heteroelements present.

The heteropolyanions are prepared in aqueous solution, usually by one of the following types of reaction:

a) The acidification of a solution containing the simple oxoanions.

$$HPO_4^{2-} + MoO_4^{2-} \xrightarrow{H^+,\ 25\ ^\circ C} PMo_{12}O_{40}^{3-}$$

$$WO_4^{2-} \xrightarrow[100\ ^\circ C]{\text{excess } H_3AsO_4} As_2W_{18}O_{62}^{6-}$$

b) The introduction of the heteroelement cation after preliminary acidification:

$$WO_4^{2-} \xrightarrow[2)\ Co^{2+},\ 100\ ^\circ C]{1)\ H^+ \text{ to } pH\ 6} Co_2W_{11}O_{40}H_2^{8-}$$

$$MoO_4^{2-} \xrightarrow[2)\ Mn^{2+},\ S_2O_8^{2-},\ 90\ ^\circ C]{1)\ H^+ \text{ to } pH\ 4.5} MnMo_9O_{32}^{6-}$$

[1]) The elements of one or more molecules of *constitutional water* may be present, as in $CrMo_6O_{24}H_6^{3-}$.

c) The controlled degradation of another heteropolyanion:

$$P_2W_{18}O_{62}^{6-} \xrightarrow{HCO_3^-,\ 25\,°C} P_2W_{17}O_{61}^{10-}$$

Many variants are possible. For procedures, see 'Inorganic Syntheses' (*24*), *Brauer* (*25*), and the references cited elsewhere in this Review. The anions are most commonly isolated as the ammonium or potassium salts. In general, the lithium and sodium salts are very soluble in water, and salts of large cations are insoluble. Solutions of the free acids of many heteropolyanions, particularly of the

$XM_{12}O_{40}{}^{(8-n)-}$, $X_2M_{18}O_{62}{}^{(16-2n)-}$, $XM_6O_{24}{}^{(12-n)-}$, $XM_6O_{24}H_6{}^{(6-n)-}$, and $XM_{12}O_{42}{}^{(12-n)-}$ series,

may be prepared by cation-exchange (*26*); in the first two cases the crystalline hydrates may be obtained by evaporation of the 'etherates' following extraction from acidified aqueous solution (*24, 25*).

2. Structures

2.1. General Comments

The structures of a number of heteropolyanions (Table 1 and Figures 1 to 6) have been determined by X-ray diffraction. Other heteropolyanions have been shown by crystal isomorphism and chemical analysis to have the same structures. Each anion contains a number of MO_6 octahedra ($M = Mo$ or W) sharing edges (Figs. 4,5), corners and edges (Figs. 1, 2, 3), or corners and faces (Fig. 6) with one another. In addition, each octahedron shares a corner with an XO_4 tetrahedron (Figs. 1, 2, 3), an edge with an XO_6 octahedron (Figs. 4, 5), or a face with an XO_{12} icosahedron (Fig. 6), X being the heteroatom. In Figs. 1 and 4 the oxygen atom positions are represented by a ball model on the same scale as the linked-polyhedron model and with the same orientation. The balls are close-packed in Fig. 1 (mixture of cubic and hexagonal) and in Fig. 4 (cubic); in fact most heteropolyanions of known structure may be regarded (*27*) as fragments of a close-packed ternary oxide with Mo or W atoms in octahedral sites and one or two heteroatoms in central tetrahedral or octahedral sites. The more open $XMo_{12}O_{42}^{4-}$ and $P_2Mo_5O_{23}^{6-}$ structures are exceptions.

Table 1. The classes of heteropolyanions of known structure. Single-crystal X-ray structure determinations (references cited) were made on the anions containing the heteroatoms in italics. For the many other heteropolyanions of unknown or uncertain structure, and for derivatives of the first two series, see text

Formula type	Figure	Central group	M = Mo	M = W	Ref.
$X^{+n}M_{12}O_{40}^{(8-n)-}$	1	XO_4	Si^{IV}, Ge^{IV}, P^{V}, As^{V}, Ti^{IV}, Zr^{IV}	B^{III}, Si^{IV}, Ge^{IV}, P^{V}, As^{V}, Al^{III}, Fe^{III}, Co^{II}, Co^{III}, Cu^{I}, Cu^{II}, Zn^{II}, (Cr^{III}, Mn^{IV}, Te^{IV}) [Ga^{III}]	*(4, 8–12)*
$X_2^{+n}M_{18}O_{62}^{(16-2n)-}$	2	XO_4	P^{V}, As^{V}	P^{V}, As^{V}	*(13)*
$X_2^{+n}Z_4^{+m}M_{18}O_{70}H_4^{(28-2n-4m)-}$	3	XO_4	—	X = P^{V}, As^{V} Z = Mn^{II}, Co^{II}, Ni^{II}, Cu^{II}, Zn^{II}	*(14)*
$X_2^{+n}M_5O_{23}^{(16-2n)-}$	—	XO_4	P^{V}	—	*(15, 16)*
$X^{+n}M_9O_{32}^{(10-n)-}$	4	XO_6	Mn^{IV}, Ni^{IV}	—	*(17)*
$X^{+n}M_6O_{24}^{(12-n)-}$	5	XO_6	Te^{VI}, I^{VII}	Ni^{IV}, Te^{VI}, I^{VII}	*(18, 19)*
$X^{+n}M_6O_{24}H_6^{(6-n)-}$	5	XO_6	Al^{III}, Cr^{III}, Co^{III}, Fe^{III}, Ga^{III}, Rh^{III}, Mn^{II}, Co^{II}, Ni^{II}, Cu^{II}, Zn^{II}	Ni^{II}	*(11, 20, 21)*
$X_2^{+n}M_{10}O_{38}H_4^{(12-2n)-}$	5	XO_6	Co^{III}	—	*(22)*
$X^{+n}M_{12}O_{42}^{(12-n)-}$	6	XO_{12}	Ce^{IV}, Th^{IV}, U^{IV}	—	*(23)*

( ); existence of anion, or membership of series, requires confirmation.
[ ]; closely-related 11-tungstate.

The X-ray scattering from a heteropolyanion is dominated by the heavy Mo or W atoms. A trial structure in which the oxygen atoms are approximately located may be suggested by the heavy atom positions and by the distances between them, which are appreciably greater if the MO_6 octahedra share corners than if they share edges or faces (Table 2). However, the oxygen positions have never been directly determined in certain cases (*13, 17, 20*). We draw attention to a recent critique (*4*) of the quality of the structure determinations up to 1971.

Table 2. Molybdenum-molybdenum and tungsten-tungsten distances in Å for different modes of linking MO_6 octahedra

Anion	Corner-sharing.	Edge-sharing	Face-sharing	Ref.
$CeMo_{12}O_{42}^{8-}$	3.82	—	3.18	(*23*)
$CrMo_6O_{24}H_6^{3-}$	—	3.33	—	(*21*)
$TeMo_6O_{24}^{6-}$	—	3.34	—	(*18*)
$P_2Mo_5O_{23}^{6-}$	3.65	3.36	—	(*15*)
$H_2P_2Mo_5O_{23}^{4-}$	3.68	3.38	—	(*16*)
$PW_{12}O_{40}^{3-}$	3.74	3.32	—	(*9*)
$GaW_{11}O_{40}H_2^{9-}$	3.74	3.32	—	(*4*)
$P_2W_{18}O_{62}^{6-}$	3.70, 3.75	3.35	—	(*13*)
$NiW_6O_{24}H_6^{4-}$	—	3.31	—	(*20*)
$NiW_6O_{24}^{8-}$	—	3.21	—	(*19*)

The inter-atomic distances have been averaged (cf. (*4*)) according to the presumed high symmetry of the isolated anions.

Before the principal types of heteropolyanions are discussed separately, three generalisations may be noted.

(i) The heteroatom is bonded to one continuous polytungstate or polymolybdate 'ligand'. Representations such as

$Si(W_3O_{10})_4^{4-}$, $Cr(MoO_4H)_6^{3-}$, or $Ce(Mo_2O_7)_6^{8-}$,

which suggest that *discrete* ligands of the indicated formulae are attached to the heteroatom, are misleading.

(ii) In all the structures discussed here, each MO_6 octahedron shares at least one atom with the central XO_n polyhedron. Moreover, no anion with an M:X ratio greater than 12:1, in which M is Mo or W, has yet been proved to exist; reports of such anions should therefore be regarded with suspicion especially when based only on solution measurements or on the analysis of a crude product. For instance, a $Te^{IV}W_{18}$ anion has been reported (*28*). The coordination number of Te(IV) is unlikely to exceed six. Eighteen WO_6 octahedra, sharing edges with one another, could be placed so that twelve each shared an edge and six each shared a corner with a central octahedron, but (a) the resultant assembly $TeW_{18}O_{44}$ would be *cationic*, and (b) appreciable distortion of the inner WO_6 octahedra to relieve W—W repulsion (see below) would not be possible. A TeW_{27} species (*29*) seems still less probable.

(iii) The structures shown in Figs. 1 to 6 are 'ideal'. In reality the MO_6 octahedra are distorted, and the oxygen positions are not exactly those of close-packed spheres. Where the oxygen positions have been precisely determined the shortest bond in each MO_6 group is to the unshared oxygen(s) at the anion periphery, and the longest bond is to an inner oxygen shared with other M atoms and with X. This is exemplified in Table 3. The inverse dependence of the bond lengths to oxygen on the coordination number of the oxygen is also found for finite isopolyanions (*18, 30*), infinite-chain polyanions (*31*), and MoO_3 (*32*). In the case of the heteropolyanions three complementary explanations have been put forward.

Table 3. Bond lengths in MoO_6 and WO_6 groups (Å). The lengths of bonds presumed to be symmetry-related in the isolated anion have been averaged (cf. (*4*)).

Anion	M—O (unshared)	M—O (M)	M—O (X, M)	Ref.
$CeMo_{12}O_{42}^{8-}$	1.68	1.98	2.28	(*23*)
$CrMo_6O_{24}H_6^{3-}$	1.71	1.94	2.29	(*21*)
$TeMo_6O_{24}^{6-}$	1.71	1.94	2.30	(*18*)
$Co_2Mo_{10}O_{38}H_4^{6-}$	1.67 to 1.73	1.89 to 1.95	2.01, 2.26 to 2.31	(*22*)
$P_2Mo_5O_{23}^{6-}$	1.72 to 1.73	1.90 to 1.93	2.18 to 2.40	(*15*)
$H_2P_2Mo_5O_{23}^{4-}$	1.70 to 1.72	1.90 to 1.96	2.20 to 2.41	(*16*)
$PW_{12}O_{40}^{3-}$	1.84	1.93, 1.97	2.29	(*9*)
$SiW_{12}O_{40}^{4-}$	1.68	1.91, 1.96	2.38	(*12*)
$GaW_{11}O_{40}H_2^{9-}$	1.72	1.91, 1.99	2.18	(*4*)
$Co_2W_{11}O_{40}H_4^{6-}$	1.71	1.93, 1.97	2.14	(*33*)
$NiW_6O_{24}^{8-}$	1.78	1.96	2.15	(*19*)

The reported range of W—O bond lengths (1.43 to 2.49 Å) in $CoW_{12}O_{40}^{5-}$ (*10, 27*) has been queried (*4*).

a) The heteropolyanions may be regarded as oxide fragments (*27*). The outer, unshared oxygens will be strongly polarised by the adjacent positive atoms. Interior oxygens, with positive atoms on all sides, are less polarised in any one direction. The net effect is that ion — induced dipole forces displace the heavy atom outwards. Also the nucleophilic character of the exterior oxygens is reduced (*27*), in agreement with the low degree of hydration in solution, and with the weakness of the anions as Bronsted bases and (with the exception of the $XM_{11}O_{39}^{(12-n)-}$ series) as ligands for cations.

b) The electronic structures of the $XM_{12}O_{40}^{(8-n)-}$, $XM_6O_{24}^{(12-n)-}$, and $XMo_9O_{32}^{6-}$ anions have been discussed in valence-bond terms (*34*). The requirement that the formal charge on each atom is to be small and that large charge separations between anion core and periphery are to be avoided results in the assignment of bond orders which vary inversely with the experimentally-determined bond lengths. For example, if the M—O (X, M) bonds are one-electron, the M—O (unshared) bonds are three-electron, and all other bonds are two-electron, then all atoms in $TeMo_6O_{24}^{6-}$ other than the unshared oxygens $\left(-\frac{1}{2}\right)$ are uncharged, while in $PW_{12}O_{40}^{3-}$ the formal charges are P, $+1$; O (central tetrahedron), $+\frac{1}{2}$; O (unshared), $-\frac{1}{2}$, other atoms, zero. Four bond lengths to oxygen are predicted for $MnMo_9O_{32}^{6-}$ (about 1.75—1.80, 1.85, 1.95, and 2.30 Å); these have yet to be verified.

c) Coulombic repulsions between the heavy atoms are lessened if the latter are off-centre in their respective octahedra (*35*). A set of three edge-linked octahedra is most stable, in this respect, if the heavy atoms define an equilateral triangle; such a grouping is particularly evident in the $XM_{12}O_{40}^{(8-n)-}$ anions.

2.2. The $XM_{12}O_{40}^{(8-n)-}$ Series; the Keggin Structure (*8*)

These anions (Fig. 1) have overall the high point symmetry (T_d) of the central XO_4 group. The M atoms lie at the corners of an almost regular cubo-octahedron; they are slightly displaced from the edge-centers of the associated cube. Although the anion is compact, it will accommodate heteroatoms (Table 1) which differ considerably in size, as is shown by the range of standard crystal radii for heterocations:

0.50 Å (Al^{3+}) to 0.96 Å (Cu^{+}),

and by the range of covalent radii for non-metal heteroatoms:

0.85 Å (B) to 1.22 Å (Ge).

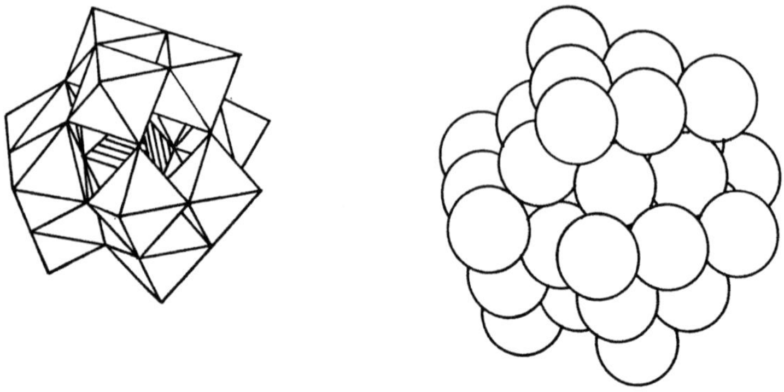

Fig. 1. The $X^{+n}M_{12}O_{40}^{(8-n)-}$ anion: the Keggin structure. Left: Linked-polyhedron representation. Right: Ball model showing oxygens atoms only: same scale

The experimental X—O distances reported to date are

Si, 1.63; P, 1.71; Ga[2]), 1.87; Co (III), 1.88 Å.

The non-occurrence of S(VI), Se(VI), and Te(VI) heteroatoms (one of which should be of the right size) in the Keggin anion has been ascribed (*34*) to the high positive formal charge that would be borne by the central XO_4 group. However, 12-tungstoselenic(IV) and -telluric(IV) acids have been reported (*36*, *37*); the latter is tetrabasic, which is consistent with a Keggin structure, but no X-ray results are available. It would be interesting to know whether the lone pair on Te(IV) is sterically active. The Keggin anion can certainly accommodate Jahn-Teller distortion (high-spin Co(III), ^5E) (*10*, *27*).

The existence of $XMo_{12}O_{40}^{4-}$ with X = Ti, Zr, and Mn(IV) has been disputed. Powder diffraction studies of their salts were reported (*38*), but the preparations could not be repeated (*39*). More recent work has verified the existence of the Ti and Zr complexes (*40*, *41*), but it seems that the only molybdomanganate(IV) anion is $MnMo_9O_{32}^{6-}$. The $XW_{12}O_{40}^{(8-n)-}$ anions with X = Co(II) and Co(III) have been thoroughly studied (section 4.1.). Other hetero-12-tungstates with metal heteroatoms (Table 1) were assigned the Keggin structure on the basis of chemical analysis and X-ray powder diffraction (*42*, *43*, *44*)[3]) but only the Fe(III) and Cu(II) complexes appear to have been re-examined (*46*, *47*). Attempts to repeat the preparation of $CrW_{12}O_{40}^{5-}$ failed (*48*); X-ray data for the 11-tungsto-

[2]) This anion is in fact a closely-related hetero-11-tungstate; see Section 2.3.
[3]) Certain of these were originally reported to be X_2W_{22} anions (*45*).

chromate actually obtained were not given, and it is uncertain whether this is the same as the anion reported earlier (*43*).

The abundant unit cell data for salts of Keggin anions and for the hydrated free acids (summarised in Ref. (*4*) and (*49*)) show that crystal isomorphism depends chiefly on similarity of degree of hydration, rather than on the cation:anion ratio or on the anion charge. Only a few modes of packing for the anions are seen in salts of monatomic cations and of ammonium; the cell is commonly cubic (either body-centred or diamond-like, with each anion inverted with respect to its eight or four neighbours, respectively) or hexagonal (with the anions in the relative position of the silicon atoms in quartz, on helices extending throughout the crystal). Other packing modes are recognisable as distorted forms of the diamond-like one. The shape of the Keggin anions permits them to interlock slightly; for instance, the distance between anion centres in $H_3PW_{12}O_{40} \cdot 29\,H_2O$ (approx. 10.10 Å) is less than twice the greatest P—O distance within the anion (5.36 Å). The voids remaining within the array of anions, and the channels penetrating the hexagonal crystals, contain the lattice water and the cations. If the anion is highly charged, or the cation is large, the preferred packing arrangement may not leave enough space for the cations, so that acid salts alone can be crystallised from solution: for example,

$K_5HCoW_{12}O_{40} \cdot 15H_2O$ (*50*), and $Cs_3HSiW_{12}O_{40} \cdot 2H_2O$, $Cs_3H_2BW_{12}O_{40} \cdot 2H_2O$, and $Cs_3H_3H_2W_{12}O_{40} \cdot 2H_2O$ (metatungstate) (*51*).

The existence of anions isomeric with the Keggin anion is now established. In the $XW_{12}O_{40}^{(8-n)-}$ series (X = B, Si, Ge) the Keggin anion (α-isomer) and the β-anion may be separated by fractional crystallisation (X = B, Si) (*52*) or may be prepared separately by varying the conditions of acidification of the mixture of simple oxoanions (X = Si, Ge) (*53*). The two compounds obtained in each case are similar in their spectra and chemical properties and may be recrystallised without change. In the Keggin anion (Fig. 1) the MO_6 octahedra form four sets of three, with edge-sharing within each set and corner-sharing between them. An isomer would be obtained by rotating one set by 60° about a three-fold axis and reattaching it (*54*, *55*), the same corners as before being unshared. Recent X-ray work shows that β—$SiW_{12}O_{40}^{4-}$ has this structure (*56*). Three more structures would be obtained by rotating the other three sets by 60° in turn (*55*)[4]). The distances between unshared

[4]) The structure in which all four sets have been rotated, and the octahedra share edges only, was proposed as an alternative to the Keggin structure (*57*), though not in connection with isomerism. It is the structure of $Al_{13}O_{40}H_{48}^{7+}$ (*58*).

oxygen atoms, and also between the heavy atoms, are greatest in the Keggin anion itself, and the net charge is spread most evenly over the anion. This structure should be the most stable one (*34*), as is indeed the case (*53*). In the more labile $XMo_{12}O_{40}^{(8-n)-}$ series (X = Si, Ge, P, As) the existence of two forms (also termed α and β) has been deduced from spectrophotometric and polarographic studies (*54, 59, 60, 61*). The stable form in aqueous solution is α (X = Si, Ge, P) and this is presumably the anion shown by X-ray powder diffraction to have the Keggin structure, though this does not seem to have been directly verified. The β-form may be stabilised in mixed solvents (conflicting results have been reported for X = As (*54, 61*); this anion is in any case unstable). The isolation of $\beta-H_4SiMo_{12}O_{40}$ and of the four-electron reduction product of $\beta-H_3$-$AsMo_{12}O_{40}$ has been reported (*60*).

2.3. Other Anions with the Keggin Structure

A number of other anions have either the Keggin structure or one closely related to it.

(i) The reduced heteropolyanions are discussed in Section 5.

(ii) X-ray powder diffraction shows that metatungstate ion is isomorphous with $PW_{12}O_{40}^{3-}$, etc. (*51, 62*). Analysis of its salts and dehydration studies (*63*) show that one molecule of constitutional water is present: $H_2W_{12}O_{40}^{6-}$. The broad-line proton magnetic resonance spectrum of its salts confirms this (*64*); the PMR spectrum in D_2O shows that the hydrogens are not exchangeable with the solvent (*65*) and presumably occupy the central tetrahedron.

(iii) A Keggin anion may contain atoms of *two* of the elements V(V), Mo(VI), W(VI).

a) A series of anions $XMo_xW_{12-x}O_{40}^{(8-n)-}$ ($X = Si, P; x = 2, 4, 6, 8, 10$) has been reported (*66*). The powder diffraction patterns of the compounds formulated

$H_3PMo_6W_6O_{40} \cdot 5H_2O$, $Cs_3PMo_6W_6O_{40} \cdot 2H_2O$,
and $Cs_3HSiMo_6W_6O_{40} \cdot 2H_2O$ (*67*)

show that the anions present are isomorphous with the limiting members of the series, but do not exclude the possibility of solid solutions containing either anions of different Mo:W ratios, or isomeric anions (see below) of a single composition.

b) The anions $PV_xM_{12-x}O_{40}^{(3+x)-}$ ($M = Mo$ or W, $x = 1, 2, 3$) (*68–72*), also $PV_4W_8O_{40}H^{6-}$ and $PV_6W_6O_{40}H_3^{6-}$ (*72*), are well established. The basicity of the acids (*68, 69, 72*) and the powder pattern of

$Cs_3H_2PV_2Mo_{10}O_{40} \cdot 2H_2O$ (*67*)

agree with the Keggin formula. The powder patterns of

$H_5PV_2M_{10}O_{40} \cdot 30H_2O$

have been interpreted (*73*) in terms of a cubic array of Keggin anions, the anions at the corners and edge-centres of the doubled unit cell being inverted with respect to the others. Other anions of partial formulae SiV_2Mo_{10}, SiV_2W_{10}, GeV_2W_{10}, AsV_2W_{10}, and SeV_2W_{10} (*74*) may have the Keggin structure.

A novel type of isomerism is predicted for these anions (*69*). Five different $PV_2Mo_{10}O_{40}^{5-}$ anions, all with the Keggin structure, are possible: one in which the V atoms have replaced Mo atoms in the *same* set of three edge-linked octahedra, and four in which the V atoms have entered *different* sets. Two of the latter will each exist in enantiomeric forms. There are eleven possible $PV_3Mo_9O_{40}^{6-}$ anions, excluding optical isomers: one of the type $P(V_3)(Mo_3)_3$, five of the type $P(V_2Mo)(VMo_2)(Mo_3)_3$, and five of the type $P(VMo_2)_3(Mo_3)$, where the parentheses enclose atoms in the same set. Four of the second type and two of the third will each exist in enantiomeric forms. The ESR spectrum of V^{4+} in photo-reduced $H_5PV_2W_{10}O_{40} \cdot 30H_2O$ has been interpreted (*75*) as indicating the $P(VW_2)_2(W_3)_2$ arrangement, but does not distinguish between the four possible structural isomers. No compound for which this type of isomerism is possible has yet been shown to be pure in this respect.

The anions $V_3W_{10}O_{40}^{5-}$, $V_4W_9O_{40}^{6-}$, and $V_5W_8O_{40}^{7-}$ have been formulated $V(V_xW_{12-x})O_{40}{}^{(3+x)-}$, that is, as derivatives of an as yet unknown 12-tungstovanadate(V) with the Keggin structure, on the basis of elemental analysis, chemical behaviour, and X-ray data (*76*).

(iv) A single heavy atom in a Keggin anion may be replaced by a variety of other atoms, particularly of transition elements. The retention of the Keggin structure in anions of the general type[5])

$X^{+n}Z^{+z}M_{11}O_{40}H_y{}^{(14-n-y-z)-}$

was first established by a single-crystal X-ray study (*77*) of

$(NH_4)_7Co^{III}Co^{II}W_{11}O_{40}H_2 \cdot 13H_2O$,

all oxygen atoms in the anion being directly located. Chemical analysis and X-ray studies have indicated the same structure for, e.g.,

$SiCo^{II}W_{11}O_{40}H_2^{6-}$, (*50, 77*), $SiCo^{II}Mo_{11}O_{40}H_2^{6-}$, $PNiW_{11}O_{40}H_2^{5-}$ (*78*), $Fe^{III}NiW_{11}O_{40}H_2^{7-}$ (*79*), and $BMn^{III}W_{11}O_{40}H_2^{6-}$ (*80*).

---

[5]) The first symbol will always denote the atom in the central tetrahedron; the second, the atom replacing one heavy atom M.

These will be considered as coordination complexes in section 4.2. We note here merely that salts of these anions with simple cations belong mainly to a limited number of isomorphous series (*77, 79–81*). A notable feature is the disorder in the crystals, causing the observed space-group to be of higher symmetry than would be expected for anions of probably, true C_s symmetry in the packing arrangement adopted. For example, the anions in

$(NH_4)_7Co^{III}Co^{II}W_{11}O_{40}H_2 \cdot 13H_2O$ (probable space-group Fm3m) (*77*)

form a cubic array with doubled unit cell, each anion being inverted with respect to its six neighbours. The Co^{II} sites, one per anion, are randomly located so that X-rays 'see' an array of $Co^{III}(Co^{II}{}_{1/12}W_{11/12})_{12}$ anions each with T_d symmetry. The elucidation of the Co^{II}—O and W—O bond lengths separately would require an ordered crystal of low symmetry (*4*). Crystals containing the anions discussed in (iii), above, may also be disordered.

(v) The $GaW_{11}O_{40}H_2^{9-}$ anion has the Keggin structure, except that one tungsten atom is absent and its unshared oxygen has been replaced by H_2O (*4*). (This anion was originally formulated as a metatungstate derivative, $H_2(GaW_{11})O_{40}^{9-}$, with gallium replacing tungsten (*77, 82*)). An aluminium analogue has been reported (*83*).

$(NH_4)_7Na_2GaW_{11}O_{40}H_2 \cdot 15H_2O$

has the cubic structure just described, the disorder now being in the vacant tungsten positions. It appears very likely, from crystal isomorphism (*81*) and from their rapid reaction with divalent cations Z^{2+} to give anions of the XZW_{11} type (*84, 85*), that 'unsaturated' heteropolyanions $XW_{11}O_{39}{}^{(12-n)-}$ (X = B, Si, Ge, P, As) also have the 'defect Keggin' structure[6]).

(vi) A large number of heteropolymolybdate complexes containing phosphorus (sometimes silicon) and a second heteroatom Z have been reported (Z = Ti, Zr, Ce, Th, Hf, Nb), the assigned stoichiometry being most commonly $PZMo_{12}$ (sometimes PZ_2Mo_{12}) (*90–93*). Some salts (Cs, pyridinium) have been isolated but not purified (*91*); analyses are sometimes incomplete and X-ray data are lacking. The bulk of the work is based on spectrophotometric studies of solutions ('Job plots') coupled with solvent extraction from acidic solution. Estimates of formation constants (*92*) do not 'prove' the assumed formulae and moreover

---

[6]) Their salts are isolated following mild base attack on the corresponding 12-anions (except when X = B) (*1, 2, 86, 87*). The anions are also formed when a solution containing the oxo-anions is made slightly acidic (*1, 2, 53, 85*). $XMo_{11}O_{39}{}^{(12-n)-}$ anions (X = Si, Ge, P) also exist (*39, 88, 89*).

postulate unacceptable dissociation fragments such as $Mo_2O_7^{2-}$. The published results do not justify assertions (*93*) that both heteroatoms are *completely* enclosed in a polymolybdate sheath. Some of the complexes, at least, may be of the $PZMo_{11}$ or PZ_2Mo_{10} types (Keggin anions); the $PNb_2Mo_{10}O_{40}^{5-}$ anion has been characterised (*94*), and the formation of $XTiW_{11}O_{40}^{(10-n)-}$ (X = B, Si, Ge, P, As), stable in acidic solution, from $XW_{11}O_{39}^{(12-n)-}$ has been reported (*84*).

(vii) Isomers of the anions described in (iii)—(v), derived from isomers of the Keggin anion, could in principle exist; see also Ref. (*55*). For example, *three* $XW_{11}O_{39}^{(12-n)-}$ anions could be obtained by removal of a tungsten atom from the β—$XW_{12}O_{40}^{(8-n)-}$ isomer, in which a single W_3 set had been rotated. Evidence for at least two forms of $XW_{11}O_{39}^{8-}$ (X = Si, Ge) has recently been given (*53*).

2.4. The $X_2M_{18}O_{62}^{(16-2n)-}$ Series

The two forms of $(NH_4)_6P_2W_{18}O_{62}$ obtained by fractional crystallisation (*95*) undoubtedly contain isomeric anions. An X-ray determination of the heavy atom positions in the B-anion (*13*) confirmed the doubled formula deduced from fused-$Na_2SO_4 \cdot 10H_2O$ cryoscopy (*87*) and showed that the anion is related to the Keggin anion as follows. Three adjacent corner-linked octahedra are removed from the Keggin anion, one from each of three edge-sharing groups, to give the half-anion shown in Fig. 2. Two

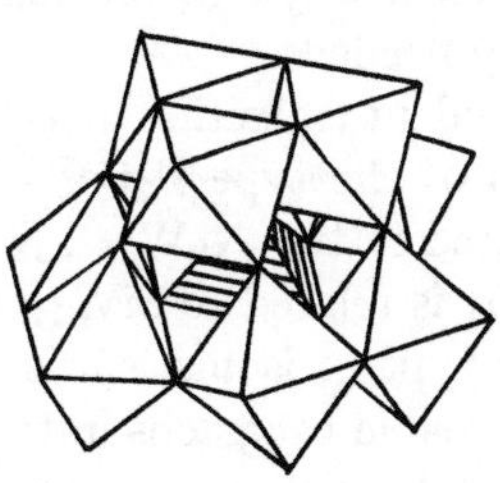

Fig. 2. The half-anion in both the A- and B-forms of $X_2^{+n}M_{18}O_{62}^{(16-2n)-}$ (linked-polyhedron representation). The six lowest vertices are shared with the other half-anion: see text

such half-anions are joined so that those vertices which were originally shared are again shared and lie in the equatorial mirror-plane of the whole anion (D_{3h}). According to unpublished work (*96*) the A-anion has a similar but centrosymmetric (D_{3d}) structure, derived from the B-structure by the rotation of one half-anion about the principal axis by 60°. Some strain is expected in the A-structure, and in fact A—$P_2W_{18}O_{62}^{6-}$

is the less stable form in solution (*98*), although interconversion does not occur in the solid state on heating (*99*). Other isomers might be derived from isomers of the Keggin anion (*55*).

Corresponding isomers of the other anions of this series (Table 1) are expected. The isolable forms of $As_2W_{18}O_{62}^{6-}$ and $P_2Mo_{18}O_{62}^{6-}$ have the B-structure (optical crystallography; see references in (*100*)). A second form of $As_2W_{18}O_{62}^{6-}$ could not be obtained (*101*). Evidence for unstable forms of the $X_2Mo_{18}O_{62}^{6-}$ ions in solution has been given (*102*).

The doubled formula $P_2V_6W_{12}O_{61}^{10-}$ has been demonstrated (*71*) for the ion formerly regarded (*70*) as $PV_3W_6O_{31}H^{5-}$. If constitutional water is present ($P_2V_6W_{12}O_{62}H_2^{10-}$) the anion could be related to either A- or B-$P_2W_{18}O_{62}^{6-}$; the substitution of 6V for 6W could be effected in many ways. The hetero-9-tungstate anions containing Be, As(III), and Sb(III) (*103*, *104*) may be members of this series or have closely-related structures; however, the true anion formulae are unknown.

In addition to the reduction products (section 5), the following anions are directly derived from the $X_2M_{18}O_{62}^{6-}$ anions:

(i) The initial products of base attack are $X_2M_{17}O_{61}^{10-}$ anions (*1*, *2*, *87*, *95*, *105*). The $X_2Mo_{17}O_{61}^{10-}$ anions are susceptible to hydrolytic degradation, but salts of $X_2W_{17}O_{61}^{10-}$ may be recrystallised. The *same* $P_2W_{17}O_{61}^{10-}$ anion is apparently obtained from both A- and B-$P_2W_{18}O_{62}^{6-}$, and is reconverted to the B-anion alone upon acidification (*95*). It is tempting to regard the $X_2M_{17}O_{61}^{10-}$ anions, by analogy with $XM_{11}O_{39}^{(12-n)-}$, simply as defect-$X_2M_{18}O_{62}^{6-}$ with one M atom and its unshared oxygen missing, but it is not then clear why only one half-anion has been affected. An X-ray study is greatly needed.

(ii) The reaction of divalent transition metal cations with $X_2W_{17}O_{61}^{10-}$ anions affords complexes of the type $P_2Co^{II}W_{17}O_{62}H_2^{8-}$ (*84*, *85*, *106*). These are, plausibly, related to the $X_2W_{18}O_{62}^{6-}$ anions in the same way that, say, $SiCoW_{11}O_{40}H_2^{6-}$ is related to $SiW_{12}O_{40}^{4-}$ (*55*, *85*, *106*). If this view is correct, a further type of isomerism is possible (*55*) because the eighteen potentially replaceable tungstens in the parent anion are not all equivalent (each half-anion (Fig. 2) contains three 'polar' and six 'equatorial' tungstens). No structural studies have yet been made apart from the measurement of some unit cell dimensions (*81*, *106*).

2.5. The $X_2^{+n}Z_4^{+m}W_{18}O_{70}H_4^{(28-2n-4m)-}$ Series

The prolonged reaction of an aqueous mixture of composition

$H^+:HPO_4^{2-}:Co^{2+}:WO_4^{2-} = 11:2:4:18$ affords the $P_2Co_4W_{18}O_{70}H_4^{10-}$

anion, whose structure (*14*) is shown in Fig. 3. A central planar group of four edge-linked CoO_6 octahedra separates two fragments of the Keggin

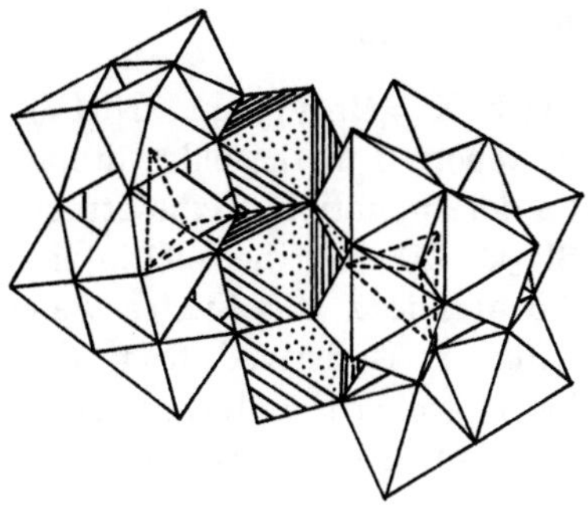

Fig. 3. The $P_2Co_4W_{18}O_{70}H_4^{10-}$ anion: linked-polyhedron representation. The cobalt atoms occupy the central group of four edge-linked octahedra

structure. The twelve octahedra ($9WO_6 + 3CoO_6$) sharing corners with each PO_4 group are linked in the manner of the β-isomer (Section 2.2.) of the Keggin anion, in which one set of three edge-linked octahedra (here, the three CoO_6) has been rotated by 60°. In the present anion, the rotation minimises crowding of the two unshared CoO_6 vertices. The anion charge as given by chemical analysis and the number of oxygens found by X-rays together require the presence of four non-acidic hydrogen atoms. The unshared CoO_6 vertices are probably occupied by water molecules. Arsenic can substitute for phosphorus, and divalent manganese, nickel, copper, and zinc for cobalt.

2.6. The $P_2Mo_5O_{23}^{6-}$ Anion

The 5-molybdodiphosphate anion is formed from HPO_4^{2-} and MoO_4^{2-} in near-neutral solution (*88, 107*). The anion in

$Na_6P_2Mo_5O_{23} \cdot 13H_2O$ (*15*) and in $Na_4H_2P_2Mo_5O_{23} \cdot 10H_2O$ (*16*)

has point symmetry close to C_2 and consists of a ring of five highly-distorted MoO_6 octahedra; adjacent octahedra share edges except that one pair only shares a corner. A PO_4 tetrahedron on either side of the ring shares three of its corners with the ring. Two oxygens on each Mo are unshared. The PO_4 group is thus unshielded, in contrast to that in 'higher' metallophosphates. The $As_2W_5O_{23}^{6-}$ anion (*108*) could have the same structure. However, chemical analysis indicates that unless constitutional water is present there are fewer than 23 oxygen atoms per empirical formula in the heteropolyanions reportedly of the types

$P_2^{III}Mo_5$ (*109*), $Sb_2^{III}W_5$ (*104, 110*),
and $S_2^{IV}Mo_5$ (*111*) and in the monomeric $Se_2^{IV}Mo_5O_{21}^{4-}$ anion (*112, 113*).

2.7. The $XMo_9O_{32}^{(10-n)-}$ Series

Only the members with X = Mn(IV) and Ni(IV) are known. The 9-molybdocobaltate(IV) reported in the older literature does not appear to exist (*114*). In the determination of the structure of rhombohedral $(NH_4)_6MnMo_9O_{32} \cdot 8H_2O$ (*17*) only the hkO intensities were considered and consequently the atomic z coordinates were not obtained. The probable structure (Fig. 4) is derived from that of a hypothetical $XMo_{12}O_{38}^{(4-n)-}$ anion in which the MoO_6 octahedra share edges with one another and with a central XO_6 octahedron. Three MoO_6 octahedra have been removed in that manner which most effectively allows distortion of the remaining structure through Mo—Mo repulsion (*35*). The resulting anion is dissymmetric (D_3) and exhibits circular dichroism (*115*).

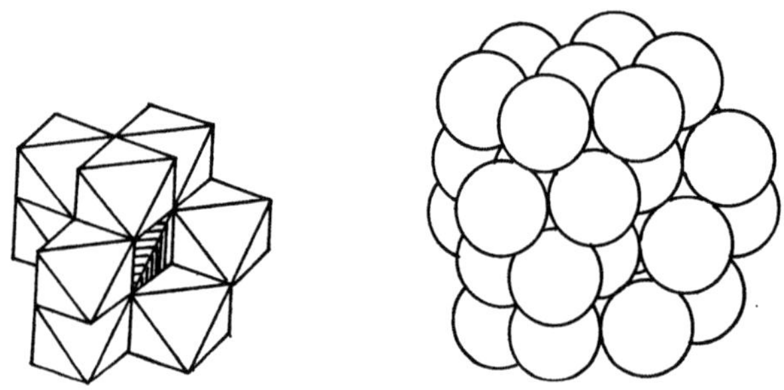

Fig. 4. The $X^{+n}Mo_9O_{32}^{(10-n)-}$ anion. Left: linked-octahedron representation. Right: Ball model showing oxygen atoms only: same scale

2.8. The $XM_6O_{24}^{(12-n)-}$ and $XM_6O_{24}H_6^{(6-n)-}$ Series

The structure shown in Fig. 5 was proposed by *Anderson* (*116*) for the $XM_6O_{24}^{(12-n)-}$ series and verified by Evans (*18, 117*) for $TeMo_6O_{24}^{6-}$. Subsequently, the $Ni^{IV}W_6O_{24}^{8-}$ anion (obtained by oxidation of

$Ni^{II}W_6O_{24}H_6^{4-}$ (*20*))

was found to have this structure (*19*). Six MO_6 octahedra share edges with a central XO_6 octahedron and with one another. The anion has D_{3d} point symmetry. The existence of anions with Te(VI) and I(VII) as heteroatoms suggests that 6-tungsto- or 6-molybdoxenate(VIII) may also be preparable.

The Anderson-Evans structure has been demonstrated for

$NiW_6O_{24}H_6^{4-}$ (*20*)

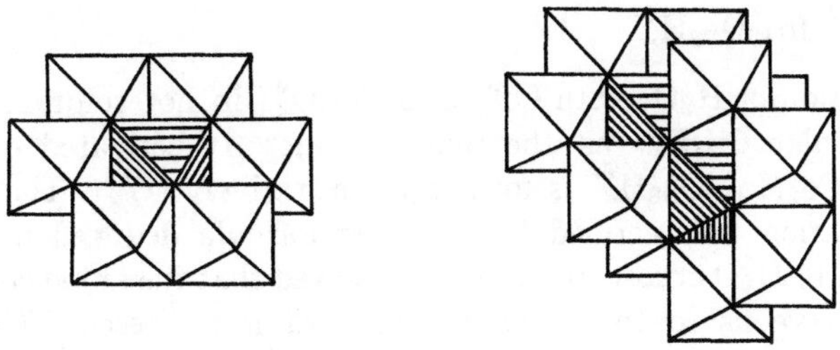

Fig. 5. Left: The $X^{+n}M_6O_{24}^{(12-n)-}$ and $X^{+n}M_6O_{24}H_6^{(6-n)-}$ anions (Anderson-Evans structure): linked-octahedron representation. Right: the $Co_2Mo_{10}O_{38}H_4^{6-}$ anion; linked-octahedron representation

(Ni and W positions only) and for $CrMo_6O_{24}H_6^{3-}$ (*21*) (Cr, Mo and O positions). Salts of these anions are isomorphous with corresponding salts of, respectively, $X^{II}Mo_6O_{24}H_6^{4-}$[7] and the other $X^{III}Mo_6O_{24}H_6^{3-}$ anions (*20, 118—120*). The X-ray work confirmed the anion weights deduced from fused-hydrate cryoscopy (*20, 114*), and also the presence of XO_6 chromophores (X = Ni(II), Cr(III), Co(III) (*20, 114, 122*)). The presence of constitutional water is required to reconcile the anion charge with the 24 oxygens in the anion, for divalent and trivalent X. See also section 3.1. Convincing arguments have been given (*20, 21*) for regarding these anions as $X(OH)_6Mo_6O_{18}^{(6-n)-}$.

A series of tungstomolybonickelates(II), $NiMo_xW_{6-x}O_{24}H_6^{4-}$, exists (*123*); it was shown that the salts obtained were not merely solid solutions of the limiting salts. Isomers are possible for $x = 2$, 3, and 4.

A number of other hetero-6-metallates of currently unknown structure have been reported. These include an $(Mn^{IV}W_6O_{24})_x^{8x-}$ anion (*124*), monomeric anions $X^{III}W_6O_{22+x}H_{2x+1}^{4-}$ (X = Al, Ga) (*83*), and a $Co^{III}W_6$ anion containing the CoO_6 group (*125*). Anions with non-metal hetero-atoms include 6-molybdoselenate(IV) (*112, 126*) and -tellurate(IV) (*127*)[8]. A reported 6-tungstophosphate(V) could not be confirmed (*87*). The 6-metallophosphate(III) anions (*105, 109*) do not seem to have been re-investigated.

[7] The anions with X(II) = Mn, Co, Cu, and Zn are unstable and their salts apparently cannot be purified by recrystallisation (*118, 121*).

[8] Ultracentrifuge measurements (*113*, 128) indicate that '6-tungstoselenate(IV)' is a tetramer, $Se_4W_{24}O_{96}H_{20}^{12-}$. The apparent formula weight of the sodium salt, 6450, is the second-highest ever reported for a heteropoly compound. Confirmation of this would be most desirable.

2.9. $Co_2^{III}Mo_{10}O_{38}H_4^{6-}$

The reaction of H_2O_2 with Co^{2+} and $Mo_7O_{24}^{6-}$ in hot solution normally affords $CoMo_6O_{24}H_6^{3-}$ as the main product. It activated charcoal is present, $Co_2Mo_{10}O_{38}H_4^{6-}$ is formed quantitatively (*114*). The structure determination (*22*) verified the doubled formula deduced from fused-$Na_2SO_4 \cdot 10\ H_2O$ cryoscopy (*114*) but showed that the anion contains 38, not 36, oxygens so that constitutional water is present. The linked-octahedron model (Fig. 5) may be obtained by removing an MoO_6 group from each of two $CoMo_6O_{24}H_6^{3-}$ ions and joining the residues so that they interlock and the CoO_6 octahedra share an edge. The anion is dissymetric (point group D_2) and has been optically resolved (*129*). The 10-tungstodigallate anion reported as $Ga_2W_{10}O_{36}^{6-}$ (*82, 83*) may have the same structure, as may a '5-tungstomanganate(IV)' (*124*).

2.10. The $XMo_{12}O_{42}^{(12-n)-}$ Series

Face-sharing between coordination polyhedra in a heteropolyanion was first detected in $CeMo_{12}O_{42}^{8-}$ (*23*). The MoO_6 octahedra form six face-sharing pairs (Fig. 6); corners are shared between octahedra in different pairs, and in addition each octahedron shares a face with the central

Fig. 6. The $X^{+n}Mo_{12}O_{42}^{(12-n)-}$ anion: linked-polyhedron representation showing face-sharing between MoO_6 octahedra in pairs

CeO_{12} icosahedron. The set of 'ideal' oxygen positions has very nearly icosahedral symmetry, but because of the presence of the Mo atoms and the marked distortion of each octahedron (*23*) the true anion symmetry is T_h. Isomorphous anions containing Th(IV) (*130*) and U(IV) (*131*) are known and the existence of anions containing Pr(IV), Np(IV), Pu(IV), and Am(IV) has been predicted (*131*). An unstable 12-molybdostannate-(IV) anion (*121*) of unknown structure has never been re-examined.

The following species are reported to be formed from $CeMo_{12}O_{42}^{8-}$:

(i) Reduction by tin(II) (*132*) or by controlled-potential electrolysis (*133*) gives 12-molybdocerate(III).

(ii) In 2 M to 8 M HNO_3, decomposition to $CeMo_{10}O_{36}^{8-}$ and $CeMo_8O_{30}^{8-}$ is said to occur (*134*).

(iii) Complexes with lanthanide cations, having Ln:Ce = 2:1, are alleged to be formed in solution. Solids believed to contain $Ln_2(OH)_2CeMo_{12}O_{42}^{4-}$ have been isolated (*135*).

2.11. Other Heteropolyanions

(i) The heteropolyanions with low M:X ratios and non-metal heteroatoms. Species of stoichiometry X_2M_5 have already been mentioned (Section 2.6.). Reported hetero-3-metallates of unknown structure include $PW_3O_{13}^{3-}$ (*87, 136*), and $AsMo_3O_{13}^{3-}$ (*137*), also an $(As^{III}W_3)_x$ acid with an *apparent* formula weight (light-scattering) of about 7000 (*138*). The formation of these heteropolyanions in solution should be particularly amenable to investigation by the potentiometric techniques that have been developed (*139*) for polynuclear and multi-component systems, provided that the protonation and polycondensation equilibria in the same medium involving the X and M elements separately have been elucidated. Computer treatment of the data is necessary, and X-ray studies of any solid phases obtained should be made. A recent study of the $H^+:HPO_4^{2-}:MoO_4^{2-}$ system (*107*) has pointed the way.

(ii) Tungstovanadates(V). In addition to the anions mentioned in section 2.3., the anions $V_xW_{6-x}O_{19}^{(2+x)-}$ exist ($n = 1,2$, and possibly 3) (*140*) and are very probably isostructural with $Nb_6O_{19}^{8-}$ and $W_6O_{19}^{2-}$.

(iii) Heteropolytungstates with lanthanide and actinide heteroatoms. Anions $XW_8O_{28}^{(8-n)-}$ (X = Ce(III), Ce(IV), U(IV)) have been reported (*141*). Attempts to repeat the preparation of the tungstocerates, however, afforded only 10-tungstocerates, $CeW_{10}O_{35+x}H_{2x}^{(10-n)-}$ (*142*). The unit cell data for $Na_6CeW_{10}O_{35+x}H_{2x} \cdot (31-x)\ H_2O$ agreed with this formula; further X-ray studies are in progress. Analogous anions with La, Pr, Nd, Sm, Eu, Ho, Er, Yb, and Y as heteroatoms exist (*142*), also 10-tungstothorate (*143*).

(iv) 11-Tungstoantimonate(III) and -bismuthate(III) (*144*). The formula $XW_{11}O_{38}H^{6-}$ is based on chemical analysis. Unless the anions are really $XW_{11}O_{39}H_3^{6-}$ with W(OH)W linkages or terminal W–OH groups they are therefore *not* isostructural with those of the $PW_{11}O_{39}^{7-}$ type, which probably have a 'defect-Keggin' structure (Section 2.3.). Nevertheless they react with transition metal cations to give complexes of the type $XCoW_{11}O_{38+x}H_{2x}^{5-}$ (*110*).

(v) Hetero-21-tungstates. The $X^{V}W_{11}O_{39}^{7-}$ anions (X = P and possibly As) undergo disproportionation in acidic solution to $XW_{12}O_{40}^{3-}$ and a second heteropolyanion (*1, 2, 87, 145*). Salts of the latter (X = P) are readily crystallised and the analyses agree well with the formula $P_2W_{21}O_{71+x}H_{2x}^{6-}$, though the hydrolytic instability of the anion precluded the measurement of its formula weight (*87*). The structure of these anions is completely unknown, though it may consist of two 'Keggin' fragments.

(vi) If the older literature is consulted (for references see *Gmelin* (*146*) and *Mellor* (*147*)) it will be apparent that yet other heteropolyanions of molybdenum and tungsten may exist. For example, the formulae

$3\,R_2O \cdot 2\,PtO_2 \cdot 8\,MoO_3 \cdot xH_2O$,
$4\,R_2O \cdot PtO_2 \cdot 10\,MoO_3 \cdot xH_2O$, etc., and
$4\,R_2O \cdot PtO_2 \cdot 10\,WO_3 \cdot xH_2O$, $6\,R_2O \cdot PtO_2 \cdot 10\,WO_3 \cdot xH_2O$, etc.

(R = alkali metal or NH_4) may represent more or less impure salts of a molybdo- and a tungsto-platinate(IV) anion. Note that varying proportions of the oxide constituents were accepted at a time when the nature of heteropolyanions was not recognised; also, that many analyses were performed on crude (sometimes unrecrystallisable) products, and that some elements in addition to oxygen may have been estimated by difference.

3. Studies of Solid Heteropoly Compounds by Means Other Than X-rays

3.1. Dehydration and Broad-Line NMR Studies

(i) *Lattice Water*

The oxygen atoms of the water in $H_3PW_{12}O_{40} \cdot 29\,H_2O$ could be assigned specific positions (*9*) in accordance with the space-group. The water oxygens in $Na_3CrMo_6O_{24}H_6 \cdot 8\,H_2O$ were directly located (*21*). Nevertheless the broad-line proton MR spectrum of $H_3PMo_{12}O_{40} \cdot 29\,H_2O$ shows that reorientation and diffusion of water is rapid above 183 K, the H_3O^+ and H_2O being indistinguishable at room temperature (*148*); the activation energy for reorientation is around 12—16 kJ (*148, 149*). Likewise, reorientation occurs in $K_3AlMo_6O_{24}H_6 \cdot 7\,H_2O$ with activation energy about 22 kJ (*150*). The similarity of the proton MR spectrum of hydrated $Sr_3(PMo_{12}O_{40})_2$ to that of a zeolite has been noted (*148*). DTA and TGA studies of hydrated Keggin acids (*151*) indicate that the lattice water is indeed zeolitic and is lost endothermally and reversibly. The situation is complicated by the fact that the loss of water may result in a change in the anion packing, for example $H_3PW_{12}O_{40} \cdot x\,H_2O$; 29 H_2O, cubic (diamond-like); 21 H_2O, orthorhomic; 14 H_2O, triclinic; 5 H_2O, cubic (body-centred). Tensimetric studies of some Keggin salts (*152, 153*) reveal both definite hydrates and phases of variable composition; the sample weight vs. ambient P_{H_2O} diagrams show regions of hysteresis (perhaps due to the slow rate of establishment of equilibrium by diffusion of H_2O within the crystal). The partial loss of lattice water from

$Na_3CrMo_6O_{24}H_6 \cdot 8\,H_2O$

gives a crystallographically distinct hydrate. Once the first water is lost, from an open channel, rearrangement of the sodium ions may occur and the escape of more water is facilitated (*21*). On the other hand, the dehydration of $K_6SiNiW_{11}O_{40}H_2 \cdot xH_2O$ has little effect on the unit cell dimensions (*154*). The contraction of the unit cell of

$(NH_4)_7Na_2GaW_{11}O_{40}H_2 \cdot 15\,H_2O$

with *increasing* temperature has been attributed (*77*) to partial collapse of the hydrogen-bonded lattice water network (analogous to the change in liquid water between 273 and 277 K), but this interpretation has been criticised (*4*).

(ii) *Constitutional Water*

The metatungstate ion was mentioned in section 2.3. The proton MR spectra of powdered salts of $X^{III}Mo_6O_{24}H_6^{3-}$ (X = Al, Cr, Fe) (*150*) clearly show a narrow line (width 4.5 G) ascribed to -OH groups, superimposed on the water signal (width 12 G). The two kinds of protons are indistinguishable at room temperature. Similar spectra were observed for the solid, hydrated, two- or four-electron reduction products of some Keggin acids

$H_{8-n}XMo_{12}O_{40}$ (X = Si, Ge, P) (*155*)

and led to the formulations $SiMo_{12}O_{34}(OH)_6 \cdot x\,H_2O$,

$GeMo_{12}O_{32}(OH)_8 \cdot xH_2O$, and $PMo_{12}O_{33}(OH)_7 \cdot xH_2O$

(that is, as hydrates of molecular acids, and *not* as hydrated oxonium salts (cf. (iii), below) of unprotonated or partly protonated anions[9])). The broad-line spectrum of hydrated potassium 11-tungstonickeloferrate(III) at 77 K (*156*) has been interpreted in terms of an anion

$FeNiW_{11}O_{37}(OH)_3^{6-}$,

[9]) Contrast the reduction of $SiW_{12}O_{40}^{4-}$ to $H_2SiW_{12}O_{40}^{6-}$, the anion of a strong acid, in acidic solution (*46*).

the unshared oxygen on Ni and two of the oxygens shared between Ni and W being protonated. The alleged coordination of oxonium cation, H_3O^+, to Ni in the related $ZnNiW_{11}$ anion (*157*) seems most improbable.

DTA and TGA studies (*20, 118, 158*) confirm the presence of constitutional water in the $CrMo_6O_{24}H_6^{3-}$, $X^{II}Mo_6O_{24}H_6^{4-}$, and $NiW_6O_{24}H_6^{4-}$ ions. In general, the loss of constitutional water follows the loss of lattice water and is more endothermic. The exact behaviour may depend on the cation, however; for instance, the loss of constitutional water can be distinguished from the loss of the last lattice water for potassium, but not for sodium, 6-molybdochromate(III) (*21, 158*). The completely dehydrated salt of an anion containing constitutional water may still be water-soluble, giving a solution with the expected spectrum; because of this, erroneous formulae have on occasion been assigned, such as $Co_2Mo_{10}O_{36}^{6-}$ instead of

$Co_2Mo_{10}O_{38}H_4^{6-}$ (*114*) and $X_2^{III}Mo_{12}O_{42}^{6-}$
instead of $XMo_6O_{24}H_6^{3-}$ (*119, 159*).

It is possible in such cases (*158*) that the anion is re-formed in solution from soluble decomposition products.

(iii) *Acidic Protons*

The isomorphism of many hydrated Keggin acids and their salts (*4, 49*) suggests that the former are oxonium salts. This has been verified by broad-line proton MR at low temperatures; for example, by comparing the spectrum of $H_3PMo_{12}O_{40} \cdot 29\,H_2O$ with that of the strontium salt (*148*). Oxonium is present in other crystalline Keggin acids (*160, 161*), in $H_6P_2W_{18}O_{62} \cdot 16\,H_2O$, and in $Na_3H_3P_2W_{18}O_{62} \cdot 17\,H_2O$ (*162*). Changes in the line shapes on dehydration of the crystals have been interpreted to mean that H_3O^+ is deformed to $H^+ . . OH_2$ ($H_3PW_{12}O_{40}$, $H_4SiW_{12}O_{40}$) (*160*) or that the anion has been protonated ($H_3PMo_{12}O_{40}$, $H_6P_2W_{18}O_{62}$) (*149, 162*). The higher hydrates of 12-molybdoceric(IV) and -thoric acids appear to be $(H_3O)_{5-6}XMo_{12}O_{39-40}(OH)_{2-3} \cdot xH_2O$ (*163*) (in solution, two of the eight hydrogens are less acidic than the others (*164*)). The proton MR spectrum of a solid, anhydrous Keggin acid or acid salt is a narrow line (width < 1 G) even at low temperatures, due to equivalent well-separated protons (*165*). It is uncertain, however, what reliance may be placed on the calculated interproton distances. The H—H distance in anhydrous $H_3PW_{12}O_{40}$, for instance, is variously given (*165*) as 4.02 and 4.50 Å. Neither distance agrees with a P—H distance (*160*) of 2.60 Å (which in any case places the protons well inside the anion) if the protons are located (*166*) on three-fold axes of the anion; each proton is assumed to be equidistant from, and presumably equally bonded to, three oxygen atoms. This novel arrangement requires confirmation.

A comment is needed on the basicities of the acids of the 'Keggin' and 18-series. The interest in coordination chemistry aroused by Werner's work led to the formulation of heteropolyanions as complexes with several discrete ligands attached to the heteroatom; thus the anions now known to have the Keggin structure, with 40 oxygens, were formulated $X(M_2O_7)_6^{(12-n)-}$. This led in turn to the supposition that a compound such as $Na_3PW_{12}O_{40} \cdot xH_2O$ was really an acid salt $Na_3H_4PW_{12}O_{42} \cdot \cdot (x-2)H_2O$, the Keggin acids in general possessing $(8-n)$ strong and four weak acid functions. Similarly $Na_6P_2W_{18}O_{62}$, for instance, was regarded as $Na_6H_6[P_2O_2(W_2O_7)_9]$. The retention of these views by some workers, despite the results of X-ray studies, was due to the preparation of what were apparently 'salts of high substitution' (*167*). It was insufficiently appreciated that degradation of the 'saturated' heteropolyanions to $XM_{11}O_{39}^{(12-n)-}$ and $X_2M_{17}O_{61}^{10-}$ ions with higher charges than their precursors may begin, in some cases, at a *p*H as low as 2 (*1*, *2*) and that the first inflection in the titration curve for the acid may represent the formation not of the neutral salt but of the 11- or 17-anion (*87*). For other work casting doubt on the salts of high substitution, see (*168*). The results of TGA and other studies of Keggin salts agree with the currently-accepted formulae (*152*, *169*). Thus hydrated $Na_3PW_{12}O_{40}$ and $Li_4SiW_{12}O_{40}$ are completely dehydrated reversibly, the binding of the last two waters being indistinguishable from that of the rest; whereas the loss of the last 0.5 H_2O from hydrated $K_3HGeW_{12}O_{40}$ and $Li_3HSiW_{12}O_{40}$ is irreversible and is accompanied by decomposition of the anion.

3.2. Infra-Red Spectra

The IR studies to date (*69*, *76*, *140*, *170—174*) have mostly been of the Keggin acids and their salts. The low-frequency region (<300 cm^{-1}) was in most cases not examined. Few Raman spectra have been recorded (*173*), and band polarisations have not been studied. In Table 4 a *partial* list of the reported frequencies and assignments for $PW_{12}O_{40}^{3-}$ and $SiW_{12}O_{40}^{4-}$ is given.

Discrepancies will be noted in:

(i) Observed frequencies. These may result from differences in the cation, from errors in instrumental calibration, and from different sample preparations (pellet or mull). The actual spectra of each anion, where published, are more similar than the table of frequencies suggests.

(ii) Assignments. These require more comment. They have been made from a knowledge of the spectra of the simple $XO_4^{(8-n)-}$ and MO_4^{2-} ions, and from observations of the changes in the spectrum of the heteropoly-

Table 4. Reported wavenumbers (cm^{-1}) and assignments for selected infra-red bands of $PW_{12}O_{40}^{3-}$ and $SiW_{12}O_{40}^{4-}$

Ref. (*170*)[a]	Ref. (*171*)	Ref. (*173*)	Ref. (*174*)
$H_3PW_{12}O_{40}$ (KBr)	$Cs_3PW_{12}O_{40}$ (KCl)	$H_3PW_{12}O_{40}$ (mull)	$H_3PW_{12}O_{40}$ (mull)
[b]	1170 Terminal W—O	[b]	1150 Asymm. P—O
1087	1075 P—O	1080 Asymm. P—O	1080 Terminal W—O
985	960 W—O	980 W—O	980 skeletal
	[b]	598 OPO bend	[b]
	520 P—O	529 W—O	525 OPO bend
$H_4SiW_{12}O_{40}$ (KBr)	$Cs_3HSiW_{12}O_{40}$ (KCl)	$H_4SiW_{12}O_{40}$ (mull)	$H_4SiW_{12}O_{40}$ (mull)
[b]	1170 Terminal W—O	[b]	[b]
1025	1000 Si—O	[b]	1005 Asymm. Si—O
[b]	960 W—O	981 W—O	970 Terminal W—O
927	[b]	923 Asymm. Si—O	[b]
	530 Si—O	536 OSiO bend and W—O	540 OSiO bend

[a]) Region below 700 cm^{-1} not examined: no assignments.
[b]) No band observed in this vicinity.

anion when either the heteroatom or the heavy atom is changed. The possibility of coupling between the vibrations of XO_4 and MO_6 groups has not been adequately considered. The normal modes of the Keggin anion (T_d) are

$$9\,A_1 + 4\,A_2 + 13\,E + 16\,T_1 + 22\,T_2;$$

the T_2 modes are IR- and Raman-active, and the A_1 and E modes are Raman-active only. Suppose a T_2 mode involves an asymmetric X—O stretch corresponding to the $\nu_3(T_2)$ normal mode of the free XO_4 group. This implies a simultaneous W—O(X, W) stretch plus W—O—W angle deformation, or a skeletal deformation of the whole anion, or a combination of the two. In general, the heteropolyanions do not fulfil the requirements (*175*) for the appearance of group frequencies in the spectrum. An exception is that the short bonds from heavy atoms to unshared oxygens may be regarded as terminal bonds to light atoms; they should have higher force constants than the other M—O bonds. Nevertheless (Table 4) there is disagreement over the characteristic frequency[10]) of the terminal bonds in the Keggin anion for a given M atom and indeed some workers (*174*) believe it should depend appreciably on the heteroatom.

To date, infrared measurements have been of most use in elucidating structural relationships between heteropolyanions. For example, the IR spectrum of $V_xW_{6-x}O_{19}^{(2+x)-}$ resembles that of $W_6O_{19}^{2-}$ (*140*), and the spectrum of $V(V_xW_{12-x})O_{40}^{(3+x)-}$ resembles that of metatungstate (*76*). The spectra of isostructural pairs are sometimes indistinguishable, for example $MnMo_9O_{32}^{6-}$ and $NiMo_9O_{32}^{6-}$ (*176*). The higher-frequency IR spectra of heteropoly compounds are sometimes informative. The protonation of ether in the etherates of Keggin acids (*173*) and the presence of H_3O^+ in the crystalline hydrates (*148, 177*) have been demonstrated. Hydrated $Na_6P_2W_{18}O_{62}$ is confirmed to be a normal salt, -OH groups being absent from the anion (*172*).

4. Anions Containing Metal Heteroatoms

4.1. One Heteroatom or Two Equivalent Heteroatoms

The anions containing metal heteroelements should be regarded, not only as well-defined ternary oxide fragments (*27*), but as finite complexes of these elements in which the ligand — the polymetallate sheath — is not an independently-stable species[11]). The heteroatoms are well separated in solid salts of anions containing a single heteroatom. Thus $CrMo_6O_{24}H_6^{3-}$, for example, contains Cr^{3+} in a situation intermediate between, say, $Cr(H_2O)_6^{3+}$ or $Cr(C_2O_4)_3^{3-}$ on the one hand and a solid solution in spinel or corundum on the other. Heteropolyanions provide opportunities for the study of metal ions in unusual oxidation states or in unusual environments. For example, few stable complexes of manganese(IV) and nickel-(IV) are known other than the $X^{IV}Mo_9O_{32}^{6-}$ anions. Cerium(IV) is ap-

[10]) Strictly the twelve terminal M—O internal coordinates of the Keggin anion contribute to *two* IR-active (T_2) modes.

[11]) The $W_{12}O_{40}^{8-}$ 'ligand' in Keggin heteropolytungstate anions requires the presence of two non-acidic interior hydrogen atoms for stability, in the absence of the heteroatom (metatungstate, Section 2.3.; the molybdenum analogue is unknown).

preciably stablised towards reduction in the $CeMo_{12}O_{42}^{8-}$ and $CeW_{10}O_{35}^{6-}$ ions as compared with Ce(IV) in 1 M H_2SO_4 (redox potentials +0.70, +1.10, +1.44 V respectively) (*142*). The $Cu^{I}W_{12}O_{40}^{7-}$ anion is the only known *discrete* complex containing the tetrahedral $Cu^{I}O_4$ group (*47*); while no *discrete* tetrahedral cobalt(III) complexes are known other than $Co^{III}W_{12}O_{40}^{5-}$ (*27, 50, 178*). Even when the heteroatom group, XO_n, is less uncommon it may sometime be most profitably studied in a heteropolyanion; thus it has been stated that in $Co^{II}W_{12}O_{40}^{6-}$ alone "can the magnetic properties of tetrahedral cobalt(II) in an oxide lattice be studied with complete confidence as to site symmetry, oxidation state, uniformity of composition, and absence of antiferromagnetic interactions" (*50*).

It was stated earlier that the Keggin anions can accommodate Jahn-Teller distortions. In $Co^{III}W_{12}O_{40}^{5-}$ all the Co—O distances are 1.88 Å, but the $Co^{III}O_4$ tetrahedron is elongated along a two-fold axis so that the three opposite pairs of edges are 2.7, 3.2, and 3.3 Å in length (*10, 27, 50*) (because of the uncertainty in the oxygen positions it is not clear whether the apparent small 'twist' is real). In the $SiW_{12}O_{40}^{4-}$ isomorph (*12*) the central group is accurately tetrahedral, and it appears that the distortion in the Co(III) complex arises from the high-spin configuration ^5E, $e^3t_2^3$ (*27, 50*). It has been suggested that the magnitude of the distortion (otherwise expected to be small) arises from the 'pull' of the W(VI) atoms on the oxygens, from the sides remote from the cobalt atom. The distortion is also manifest in the splitting, by nearly 3000 cm^{-1}, of the principal ligand-field band at 8550 cm^{-1} ($^5T_2 \leftarrow {}^5E$ in T_d symmetry). The spectrum of 'tetrahedral' Co(III) in garnets (actual site symmetry S_4) is similar (*179*); in both cases the splitting is much too large to be accounted for by spin-orbit coupling. The optical and ESR spectra of $Cu^{II}W_{12}O_{40}^{6-}$ (*47*) also reveal the expected axial distortion of $Cu^{II}O_4$ ($^2T_2$, $e^4t_2^5$). In contrast the $Co^{II}O_4$ group in $CoW_{12}O_{40}^{6-}$ (ground state $^4A_2$, $e^4t_2^3$) appears to be undistorted, as expected (*50*). Since the $e^2t_2^1$ configuration ($^4T_1$) is expected to be Jahn-Teller active (provided the degeneracy is not removed by spin-orbit coupling) and since there are in any case so few examples of chromium(III) and manganese(IV) in non-octahedral sites, particular interest would attach to the confirmation and improved syntheses of the $Cr^{III}W_{12}O_{40}^{5-}$ and $MnW_{12}O_{40}^{4-}$ anions (*43, 44*).

Some other features of transition-metal heteroatoms in heteropolyanions are now considered.

(i) *Magnetic properties.* Because of the large formula weights of heteropolyanions, the diamagnetic part of the anion susceptibility is rather large and should preferably be estimated from the susceptibility of a salt of an isostructural anion which lacks paramagnetic atoms. Some older measurements (*180*) on anions with transition-metal heteroatoms in near-octahedral sites were made at room temperature only and diamagnetic corrections were estimated via the susceptibilities of MoO_3 and WO_3. Consequently the deviations of μ_{eff} from spin-only values are not significant, though the work serves to confirm the low-spin character of Co(III) in $CoMo_6O_{24}H_6^{3-}$ and $Co_2Mo_{10}O_{38}H_4^{6-}$ and the oxidation state of Mn(IV) and Ni(IV) in $XMo_9O_{32}^{6-}$. In contrast, in a careful study of the $Co^{II}W_{12}O_{40}^{6-}$ and $Co^{III}W_{12}O_{40}^{5-}$ anions between 80 and 300 K *(50)*, the diamagnetic corrections were estimated from the experimental susceptibility of $K_4SiW_{12}O_{40} \cdot 11\,H_2O$, and temperature-independent paramagnetism was also allowed for. Curie-Weiss behaviour was observed in both cases with θ equal to -0.1 K for Co(II) and $+1.1$ K for Co(III) (convention: $\chi = C/(T+\theta)$), confirming the expected magnetic dilution of the heteroatoms[12]) and also the regularity of the $Co^{II}O_4$ tetrahedron and the axial elongation of $Co^{III}O_4$. The values of μ_{eff} were 4.27 B.M. for Co(II) and 5.09 B.M. for Co(III); the former required that λ was reduced to 70% of its free-ion value, indicating appreciable *d*-electron delocalisation.

(ii) *Electronic spectra.* Some of the expected *d–d* bands may be obscured by intense absorption due to other causes, which will be discussed first. All heteropolyanions absorb strongly in the ultra-violet with ε_{max} in excess of 10^4. The position of the UV absorption edge depends on the anion type, but in general ε will exceed 10^3 beyond 29—30,000 cm^{-1} in tungstates, and 25—27,000 cm^{-1} in molybdates. Thus the $^3T_{1g}(P) \leftarrow {}^3A_{2g}$ transition is observed at 26,600 cm^{-1} in the spectrum of $Ni^{II}W_6O_{24}H_6^{4-}$ *(20)* but is obscured in that of $Ni^{II}Mo_6O_{24}H_6^{4-}$ *(114)*. The UV absorption is almost certainly due to charge-transfer transitions, $O \rightarrow Mo$ or $O \rightarrow W$. In a recent treatment *(181)*, heteropolyanions with a single unshared oxygen per heavy atom have been considered along with other oxo-complexes of the type MOX_5^{x-} having C_{4v} symmetry, where M is V, Nb, Mo, or W with a d° or d^1 configuration and X is oxygen or halogen. The wavenumbers of the lowest-energy charge-transfer (LCT) bands have been shown to fit an expression of the type

$$E = A - (B - Cq)$$

where A is characteristic of X and B, C are characteristic of the d^q metal atom. Optical electronegativity values of 2.5 and 2.3 have been assigned to Mo(VI) and W(VI) respectively in oxo-compounds. However, it is still uncertain whether the LCT transition in a heteropolyanion originates on a terminal or on a shared oxygen.

---

12) In ESR work on heteropolyanions, however, it has been found necessary to use a diamagnetic host crystal in order to resolve the hyperfine structure; see (iii), below.

In the case of certain oxidisable metal heteroatoms strong bands are observed at lower energies, which are absent from the spectra of isomorphous anions containing non-metal heteroatoms, and which cannot be assigned to d–d transitions: for example at 28,800 cm^{-1} (ε 480) for $Ce^{III}W_{10}O_{35}^{7-}$ (*182*) and at 20,600 cm^{-1} (ε 2400) for $Cu^{I}W_{12}O_{40}^{7-}$ (*47*). Since a number of classes of heteropolyanions are known to undergo reduction of the intact polymetallate skeleton, affording 'blues' (Section 5), and since in the one- and two-electron blues the added electrons are trapped on specific heavy atoms, the above bands have been assigned to 'metal-to-ligand charge-transfer', and are probably most realistically described as 'heteronuclear intervalence charge-transfer', for example

$Cu^{I}W_{12}^{VI}O_{40}^{7-} \rightarrow Cu^{II}W^{V}W_{11}^{VI}O_{40}^{7-}$ (*47*).

In other cases, where the heteroatom has an accessible lower oxidation state, what are probably oxygen-to-metal charge-transfer bands are observed; for example, in the spectrum of $Mn^{IV}Mo_9O_{32}^{6-}$ (discussed below) and $Ni^{IV}Mo_9O_{32}^{6-}$ (*183*). A band near 25,000 cm^{-1} with $\varepsilon = 1250$ in the spectrum of $Co^{III}W_{12}O_{40}^{5-}$ (*50*) may be of this type, though it has also been suggested that it represents a $4s \leftarrow 3d$ transition, with considerable delocalisation of the excited electron (*50*). Oxygen-to-cobalt charge transfer probably also accounts for the strong absorption ($\varepsilon \sim 10^3$) near 27,000 cm^{-1} by the μ-dihydroxodicobalt(III) complex

$Co_2Mo_{10}O_{38}H_4^{6-}$;

the tail of this band extends through the visible and permits only the first d–d band, ${}^4T_{1g} \leftarrow {}^4A_{1g}$ at 16,500 cm^{-1}, to be seen (*114, 184*). The extreme pleochroism of $(NH_4)_6Co_2Mo_{10}O_{38}H_4 \cdot 7\,H_2O$ has been remarked on (*22*); the polarised single-crystal spectrum should be of interest.

Virtually all measurements of d–d spectra of heteroatoms have been made on solutions at room temperature (*20, 42–44, 50, 114, 183–186*), and no transmission data for single crystals or glasses at low temperatures are available. Moreover, some expected transitions are obscured for the reasons explained earlier or because they lie well into the infra-red (*e.g.* the ${}^4T_2 \leftarrow {}^4A_2$ band of $Co^{II}W_{12}O_{40}^{6-}$ (*50*)) in a region where absorption by solvent interferes. Therefore, even when the chromophore is known to have symmetry close to O_h or T_d, the desirable excess of data over the number of parameters to be ascertained is not always achieved and it may be necessary to assume that the Racah B parameter is invariant for all states of a given complex or that the free ion C/B ratio is applicable. In addition to all this, we are dealing with several structural classes of anions; not surprisingly, therefore, we have *not* found it possible to fit closely our estimates of Δ and B_{35} from published frequencies of spin-allowed bands to the Eq. (*187*)

Table 5. Ligand field and Racah parameters (cm^{-1}) for some metal ions in heteropolyanions and other oxygen environments. Reference numbers in parentheses

Octahedral coordination								
Ni(II)	Δ	B_{35}	Cr(III)	Δ	B_{35}	Co(III)	Δ	B_{35}
$NiMo_6O_{24}H_6^{4-}$	9520	— (*114*)	$CrMo_6O_{24}H_6^{3-}$	18300	650[a] (*114, 184*)	$CoMo_6O_{24}H_6^{3-}$	17300[a]	590[a] (*114, 184*)
$NiW_6O_{24}H_6^{4-}$	9800	875[a] (*20*)						
$Ni(H_2O)_6^{2+}$	8500	930[a] (*188*)	$Cr(H_2O)_6^{3+}$	17400	730[a] (*188*)	$Co(H_2O)_6^{3+}$	17230[a]	595[a] (*188*)
$Ni^{2+}:MgO$	8600	860[a] (*188*)	$Cr^{3+}:Al_2O_3$	18100	610[a] (*188*)	$Co^{3+}:Al_2O_3$	16100[a]	570[a] (*188*)
$Ni^{2+}:MgWO_4$	7100	[b] (*189*)	$Cr^{3+}:MgAl_2O_4$	18250	700 (*190*)			

Tetrahedral coordination		
Co(II)	Δ	B_{35}
$CoW_{12}O_{40}^{6-}$	4840	672 (*50*)
$Co(H_2O)_4^{2+}$	4140[c]	820[a] (*188*)
$Co^{2+}:ZnO$	3900	700 (*191*)
$Co^{2+}:ZnAl_2O_4$	4300	730[a] (*192*)

[a] Estimated from band wavenumbers in literature.
[b] Stated (*189*) to be lower than in MgO.
[c] 4/9 of Δ value for hexaaquo-ion (*188*)[a].

$\Delta = f(\text{ligand}) \cdot g(\text{metal})$ and $(1 - B_{35}/B_{\text{free ion}}) = h(\text{ligand}) \cdot k(\text{metal})$. In general, though, Δ values are somewhat greater than the values either for the aquo-ions or for the cation in a host oxide. Values of B_{35} are generally appreciably reduced compared with the aquo-ion and are closer to those found for the cation in a host oxide. Representative values are given in Table 5. Note that the site available in a continuous oxide structure should be less adaptable to the size of the guest cation than is the central site in a heteropolyanion such as the Keggin anion (Section 2.2.). Also, in both kinds of structure the bonding to the hetero-atom or guest cation is affected by second-nearest neighbours (M(VI) or host oxide cation); but in the heteropolyanion the increased polarisation of the oxygens of the XO_n group by M^{6+} (as compared with, say, Al^{3+}), which would tend to make the X—O bond less covalent, is offset by the rather long M—O distance (Section 2.1. and Table 3). These points have been discussed in more detail elsewhere (*50*) in connection with the Keggin 12-tungstocobaltate anions.

The nephelauxetic effect is smaller for the lanthanide heteropoly-tungstate anions, $Ln^{III}W_{10}O_{35}^{7-}$, than for Ln^{3+} in, for instance, Ln_2O_3 or LnOCl (*182*). 'Hypersensitive' transitions (with abnormally large intensity) have been observed for Pr^{3+} for the first time in $PrW_{10}O_{35}^{7-}$ and also in the complexes $Pr(XW_{11}O_{39})_2^{n-}$ with 'unsaturated' hetero-polyanions. These and related Ln^{3+} (especially Eu^{3+} and Ho^{3+}) complexes have been the basis for studies of f—f transition intensities (*193*); it was concluded, for example, that for *molecular* complexes in solution or in the solid state the f—f transitions gain intensity mostly via a static perturbation, and that the 'inhomogeneous dielectric' and 'covalency' theories of hypersensitivity are not valid.

The polarised single-crystal spectra of anions where an axial component of the ligand field is expected ($XM_6O_{24}H_6^{(6-n)-}$, $XMo_9O_{32}^{6-}$) have not been studied (except in the case of $MnMo_9O_{32}^{6-}$ in the 14—16,000 cm^{-1} region (*115*)). There have been very few measurements of emission spectra (*115, 194*) or of circular dichroism (*115, 129*), and no magnetic CD studies. Measurements of these kinds may require the revision of conclusions based on solution absorption spectra. Some work on

$Mn^{IV}Mo_9O_{32}^{6-}$

illustrates this. The absorption and excitation spectra (*183, 194*) indicate that the ${}^4T_{2g}$ and ${}^4T_{1g}(P)$ states lie at 21,300 cm^{-1} and at about 25,500 cm^{-1} above the ${}^4A_{2g}$ ground state, the former being split by 700—800 cm^{-1} (the true site symmetry of Mn is D_3). The ${}^2E_g$ state lies near 14,300 cm^{-1}. B_{35} and B_{55} were estimated at about 400 and 500 cm^{-1} respectively. However, the CD spectrum of the ammonium salt in a KCl pellet[13]) reveals the E components of the T states at 20,250 and 22,950 cm^{-1} (*115*) and suggests that the peaks formerly assigned to the T states may result from $O(p\pi) \rightarrow Mn(t_{2g})$ charge transfer, from which the d–d transitions gain intensity via mixing of the excited states. The *polarised* excitation spectrum (*115*) for the $^2E_g \rightarrow {}^4A_{2g}$ phosphorescence agrees with these conclusions. The 0,0 transition was identified in the emission spectrum at 14,300 cm^{-1}. The reasonable value of 680 cm^{-1} for B_{55} was deduced.

The circular dichroism spectrum of $Co_2^{III}Mo_{10}O_{38}H_4^{6-}$ has been reported (*129*). No conclusions were drawn, except that the strong bands in the near UV were assigned to charge-transfer in the $Co(OH)_2Co$ group. The study of the circular dichroism of Keggin anions with transition-metal heteroatoms in a $K_4SiW_{12}O_{40} \cdot 17\,H_2O$ host crystal (or in the isomorphous tungstoborate) may prove rewarding. Anions in this crystal (*12*) approach very closely to the ideal T_d symmetry expected for the free anion. However, the space group ($P6_222$) contains only screw and pure rotation symmetry operations, the anions being arranged along helices. There is then the possibility, of some theoretical interest, of optical activity due to the helicity of the crystal rather than to the inherent dissymmetry of the chromophore (*cf.* α–$Ni(H_2O)_6SO_4$ (*195*)).

(iii) *Electron spin resonance.* There have been two measurements of the ESR spectrum of $MnMo_9O_{32}^{6-}$ in diamagnetic $(NH_4)_6NiMo_9O_{32} \cdot 8\,H_2O$ (*196, 197*). The hyperfine structure was resolved at the moderate dilution of 1:16 (*197*). In both cases the spectrum could be interpreted using the usual Hamiltonian for $S = 3/2$ in a ligand field of axial symmetry:

$$\hat{H} = \beta H\,(g_{\parallel} \cos\theta\, \hat{S}_z + g_{\perp} \sin\theta\, \hat{S}_x) + D\left(\hat{S}_z^2 - \frac{1}{3}\hat{S}^2\right) + A_{\parallel}\, \hat{I}_z \hat{S}_z + A_{\perp}(\hat{I}_x \hat{S}_x + \hat{I}_y \hat{S}_y)$$

and concordant results were obtained. The zero-field splitting, D, was large (0.69 cm^{-1} (*196*) and 0.86 cm^{-1} (*197*)). To reproduce the latter value by calculation required that the t_{2g} orbital set was split by 7,000 cm^{-1} (to $a_1 + e$, in D_3 symmetry)[14]).

The ESR spectrum of tetrahedral, high-spin Fe^{3+} in $FeW_{12}O_{40}^{5-}$ dissolved in $K_5BW_{12}O_{40} \cdot xH_2O$ has been obtained (*199*). The results were fittet to a spin Hamiltonian containing S^4 terms (*200*), and the

[13]) The anion racemises rapidly in solution (*183*). Enantiomeric crystals may, however, be selected by hand.

[14]) X-ray reveal a considerable twist of the MnO_6 group (*17*) but the extent of axial elongation or compression is not known. The quadrupole splitting of the $I = 5/2$ excited state in the ^{61}Ni Mössbauer spectrum of $(NH_4)_6NiMo_9O_{32} \cdot 8\,H_2O$ (*198*) also indicates a non-cubic site.

parameters obtained were compared with those for tetrahedral Fe^{3+} in garnets. The values of D and F for $FeW_{12}O_{40}^{5-}$ were much smaller, indicating less axial distortion, but *a* was of similar magnitude; a value of 1.83 Å was estimated for the Fe—O distance. Non-zero values of D and E revealed a small rhombic distortion of uncertain origin.

4.2. Anions Containing Two Heteroelements

In this section are included the heteropolyanions containing two atoms of a single, metallic, heteroelement in non-equivalent sites.

(i) The majority of the anions covered by this section are the types $X^{+n}Z^{+z}M_{11}O_{40}H_y{}^{(14-n-y-z)-}$, where M is Mo or W, and

$X_2^{+5}Z^{+z}W_{17}O_{62}H_y{}^{(12-y-z)-}$.

These were briefly mentioned in Sections 2.3. and 2.4. as derivatives of the Keggin and 18-series, respectively. The X atom occupies an inner tetrahedral site; Z has replaced a heavy atom. Following the recognition (*77*) of the first members of the XZM_{11} series (including some anions earlier regarded (*50, 201*) as dihetero-12-tungstates) a large number of anions of this and of the X_2ZW_{17} series have been reported, although full preparative and analytical details have not always been given. Tables 6 and 7 contain the chief references to the XZM_{11} anions. For the X_2ZW_{17} series, see Refs. (*80, 81, 84, 85, 106, 206, 219*). Chemical analysis, in conjunction with the requirement that the XZM_{11} and X_2ZW_{17} anions contain 40 and 62 oxygen atoms respectively, requires that most anions contain one or more non-acidic hydrogen atoms; most commonly two hydrogens are present, although there may apparently be up to four (*e.g.* $Co^{II}Co^{II}W_{11}O_{40}H_4^{6-}$, (*33*)) while in some cases the anion is unprotonated (*e.g.* Z = V(IV) (*203, 204*))[15].

In effect, a pentadentate $XM_{11}O_{39}{}^{(12-n)-}$ or $X_2W_{17}O_{61}^{10-}$ ligand is attached to the Z atom, whose sixth (unshared) coordination site is occupied by O^{2-}, OH^-, or H_2O. Oxygen atoms shared between Z and W may also be protonated (*156*). The probable true site symmetry of Z in the XZM_{11} anions is C_s, but approximates to C_{4v}. An axial component of the ligand field is apparent in the spectra of complexes with Z = Mn(III)

[15]) The apparent number of hydrogens is subject to the inevitable small errors in the analyses. It is chiefly dependent on the nature of the Z heteroatom. Thus most anions with Z = Ni(II) (*78, 106, 213, 218*) are best formulated with two hydrogens. The much-studied $Fe^{III}NiW_{11}$ and $ZnNiW_{11}$ anions (*79, 156, 157, 214—217, 221*) are stated to contain *three*, but the analyses on which this is based have not been published. The preparation of salts of the latter, and of the $ZnCo^{II}W_{11}$ anion, free from cationic Z^{2+} is difficult (*176, 222*).

Table 6. References to $X^{+n}Z^{+z}M_{11}O_{40}H_y^{(14-n-y-z)-}$ anions in which tetrahedrally-coordinated X^{+n} has the d° or d^{10} configuration. For the general reaction of Z^{+z} (including Ti(IV), V(V), and Mo(VI)) with 'unsaturated' heteropolyanions $XM_{11}O_{39}^{(12-n)-}$, see Ref. (*84*) and (*85*)

	M = W[a, c]	M = Mo[b, c]
Z = Al(III)	(*202, 203*)	—
V(IV)	(*85, 203, 204*)	(*205*)
Cr(III)	(*202, 203, 206*)	(*207*)
Mn(II), Mn(III)	(*80*)	(*80, 208*)
Fe(II), Fe(III)	(*85, 202, 209*)	(*209—211*)
Co(II), Co(III)	(*50, 55, 77, 78, 203, 212, 213*)	(*78, 208*)
Ni(II)	(*78, 154, 157, 214—217*)	(*208, 218*)
Cu(II)	(*80*)	(*80, 208*)
Zn(II)	(*213, 214*)	(*208*)
Ga(III)	(*203*)	—
Re(IV), Re(VI)	(*219*)	—

[a] X = B, Si, Ge, P, As, H_2 (metatungstate derivative), Zn, Ga.
[b] X = Si, Ge, P.
[c] Not all possible combinations of X and Z have been reported.

Table 7. References to $X^{+n}Z^{+z}W_{11}O_{40}H_y^{(14-n-y-z)-}$ anions in which both heteroatoms have partly-filled 3d subshells

	X (tetrahedral coordination):		
	Fe(III)	Co(II)	Co(III)
Z (octahedral coordination):			
Fe(III)	—	(*220*)	(*220*)
Co(II)	(*220*)	(*33, 50, 77, 201, 220*)	(*50, 77, 201, 220*)
Co(III)	(*220*)	—	—
Ni(II)	(*79, 156, 214, 221*)	—	—

and Cu(II), where a large Jahn-Teller distortion of an octahedral ZO_6 group is in any case expected (the stabilization of the +3 oxidation state for Mn in these complexes is notable (*80, 106*)). The solution spectra of anions containing Co(II) and Co(III) (*50, 55, 78, 85, 106, 203, 213*), Ni(II) (*78, 85, 106, 221*), and Cr(III) (*206*) as Z atoms are interpretable in terms of octahedral chromophores, though the bands are several times as intense as in the spectra of, *e.g.*, the centrosymmetric

$XMo_6O_{24}H_6^{(6-n)-}$

anions (X = Cr(III), Co(III), Ni(II)). The ligand field is *weaker* in these complexes than in the aquo-ions; for instance, Δ is 8600 cm^{-1} for $SiCo^{II}W_{11}O_{40}H_2^{6-}$ (*50*) and 8100 cm^{-1} for $SiNiW_{11}O_{40}H_2^{6-}$ (*78*), compared with about 9300 and 8600 cm^{-1} for the respective aquo-ions[16]. Some salts of $XCo^{II}W_{11}$ anions (in particular

$(R_3NH)_5PCoW_{11}O_{40}H_2 \cdot xH_2O$ (R = CH_3, C_2H_5))

are markedly pleochroic, and their diffuse reflectance *d—d* spectra differ considerably from solution spectra of the same anions (*213*). This has been interpreted in terms of a preferred orientation of the anions in the crystal (in contrast to the disorder more generally prevailing in salts of anions of this type; see Section 2.3.), with an accompanying distortion of the chromophore through crystal packing forces. The structures of the WO_6 and CoO_6 groups may prove to be separately determinable in such salts.

The anions containing V(IV) are of interest (*204*) as analogues of the one-electron 'blues' (Section 5). For instance, $PV^{IV}W_{11}O_{40}^{5-}$ is isoelectronic with $PW^{V}W_{11}^{VI}O_{40}^{4-}$. The optical and ESR spectra of the former anion have been studied (*204*). The usual equations relating optical and ESR parameters did not give consistent results, possibly (*204*) because of mixing-in of an excited state (20,000 cm^{-1} above the ground state) arising from heternuclear intervalence charge-transfer absorption, V(IV) → W(VI). IVCT transitions dominate the visible spectra of this and related (*85, 203, 204*) V(IV) complexes, and probably account for the bands which appear as shoulders on the UV absorption edge in the spectra of related anions containing Co(II) (*50, 78, 106*), Mn(II) (*80, 106*), and Cu(II) (*80*) and are not observed in the case of isostructural $XZnW_{11}$ and X_2ZnW_{17} complexes. Strong absorption by

---

[16]) This fact, together with the observation that ligand field strengths are not greatly different in isostructural heteropolymolybdates and tungstates and may indeed be smaller in the former (*e.g.* the $NiM_6O_{24}H_6^{4-}$ anions (*20, 114*)), casts doubt on the conclusion (based on the ESR spectra of powdered salts (*209*)) that $SiFe^{III}Mo_{11}O_{40}H^{6-}$ is a low-spin complex although $SiFe^{III}W_{11}O_{40}H^{6-}$ is high-spin. The actual magnetic moments have not been measured.

the $Co^{III}Co^{II}W_{11}$ anions throughout the visible region (*50*) probably results from homonuclear IVCT (Co(II) → Co(III))[17].

The presence of an actual water molecule coordinated to Z, when the anion contains two hydrogen atoms, was first demonstrated (*55*) for the $SiCo^{II}W_{11}$ and $SiCo^{III}W_{11}$ anions. The water in the first was shown to be replaceable by pyridine, and in the second by pyridine and ammonia. Crystalline salts of the product anions were isolated. The phenomenon seems to be a general one for XZW_{11} anions containing Co(II) (*55, 213*), Co(III) (*55, 203, 213*), Ni(II) (*55, 213*), and Cr(III) (*222*). As predicted (*55*), it has also been observed for the X_2CoW_{17} anions (*176*). These anions and their substitution products therefore constitute a class of hybrid complexes, intermediate between 'normal' heteropolyanions with metal heteroatoms and conventional coordination complexes, and the significance of this has been discussed elsewhere (*55*).

Susceptibility measurements on powders at temperatures down to 77 K shown that XZW_{11} salts in which X is a non-metal are magnetically dilute, with small or zero Weiss constants. There is no evidence for magnetic coupling between the 'outer' Z heteroatoms via intervening lattice water[18]. Consequently the magnetic behaviour of complexes in which both X and Z have partly-filled *d* subshells is of interest, since superexchange is possible *within* the anion (X and Z share an oxygen atom) but is not expected to occur *between* anions. Moreover in some cases the isomer of an anion exists, in which the 'octahedral' and 'tetrahedral' atoms have exchanged sites. Following the observation (*50, 77*) that the susceptibility of $K_7Co^{III}Co^{II}W_{11}O_{40}H_2 \cdot 12\,H_2O$ changes by only 13% between 130 and 300 K, salts of the following anions have been studied (220) ('tetrahedral' atoms given first in each partial formula):

$Co^{II}Fe^{III}W_{11}$, $Fe^{III}Co^{II}W_{11}$, $Co^{III}Fe^{III}W_{11}$,
$Fe^{III}Co^{III}W_{11}$, $Co^{III}Co^{II}W_{11}$, $Co^{II}Co^{II}W_{11}$.

The measured susceptibilities between 2 K and 300 K agreed with the values calculated from the simple Hamiltonian

$$\hat{H} = \beta H.\,(g_1\hat{S}_1 + g_2\hat{S}_2) - J\hat{S}_1 \cdot \hat{S}_2$$

[17]) This complex belongs to Class I in *Robin* and *Day's* classification of mixed-valence compounds (*223*). Unambiguous IVCT bands in the spectra of Class I compounds are apparently rare. However, the visible spectrum of this anion cannot be reproduced by summing the spectra of the $Co^{III}W_{12}O_{40}^{5-}$ and (say) $SiCo^{II}W_{11}O_{40}H_2^{6-}$ anions, although the *d—d* band of the $Co^{III}O_4$ group is observed near 8500 cm^{-1} (*50*).

[18]) With an improved diamagnetic correction (cf. Section 4.1.) to the published susceptibility data (*215*), the Weiss constant of $(NH_4)_7ZnNiW_{11}O_{40}H_3 \cdot 15H_2O$ is reduced from 32 K to 13—14 K (*176*).

where the g and S values were obtained from the spectral and magnetic properties of isostructural complexes containing a single paramagnetic atom. The values of J/k ranged from − 6 K ($Co^{II}Co^{II}W_{11}$) to − 72 K ($Co^{III}Co^{II}W_{11}$). The $Fe^{III}Co^{III}W_{11}$ anion, containing octahedral, presumed diamagnetic Co(III), obeyed a simple Curie law when impurity Co(II) was allowed for. The predicted maximum in the susceptibility of the $Co^{II}Co^{II}W_{11}$ anion ($S_1 = S_2$) at very low temperatures was not observed. The other anions, in agreement with theory, obeyed a Curie law characteristic of $|S_1 - S_2|$ at low temperatures, and approached asymptotically a second Curie law characteristic of the uncoupled spins at high temperatures. It is to be hoped that further studies of this type will be made.

(ii) Some other heteropolyanions each containing one non-metal and one transition metal heteroelement exist. The recently-characterised $P_2Co_4W_{18}O_{70}H_4^{10-}$ anion (*14*) (Section 2.5.) contains an unusual grouping of four coplanar cobalt(II) atoms (Fig. 3). The magnetic interactions in this anion and in its isomorphs containing nickel(II), etc., should be of interest. Anions of this series containing, *e.g.*, both cobalt(II) and cobalt(II), or both cobalt and nickel, may prove to be preparable. The anion of empirical formula $GeFe_2^{III}Mo_8O_{31}^{4-}$ (*210*) could possibly be dimeric, with a similar structure. The $SiCu^{II}Mo_8O_{30}^{6-}$ anion (*208*), however, is monomeric (ultracentrifuge (*208*)), though its structure is currently unknown. The structures of the anions of the type $Sb^{III}Co^{II}W_{11}O_{38+x}H_{2x}^{5-}$ (*110*) are also unknown, though the spectra indicate an octahedral site for Co, etc.

5. The Reduced Heteropolyanions ('Heteropoly Blues')

The ability of heteropolyanions of the Keggin and 18-series to undergo reduction to mixed-valence species without change of structure has long been known (*1, 2*). We shall refer to the reduced anions as 'blues' in recognition of their most striking feature, which results from strong absorption in the red with ε_{max} in the range 2,000 to 5,000 per added electron (*97, 100, 224, 225, 226*). It has recently been pointed out (*227*) that the formation of blues without change of structure is restricted, as far as is known, to the above types of heteropolyanions and to the isopolyanions $V_{10}O_{28}^{6-}$ and $M_6O_{19}^{x-}$ (including M = Mo + V, W + V); a feature of all these species is the presence of MO_6 groups containing a single unshared oxygen. Other hetero- and iso-polyanions, in which all MO_6 groups have two (cis) unshared oxygens, do not undergo reduction with the anion structure intact. An analogy has been drawn (*227*) with the existence of mononuclear complexes MOL_5, but not cis-MO_2L_4, in which M has the d^1 configuration (the comparability of the UV charge-transfer spectra of polyoxoanions and mononuclear MOL_5 oxo-complexes (*181*) was mentioned in Section 4.1).

The extended reduction of some Keggin anions has been observed. The reduction of $SiW_{12}O_{40}^{4-}$ to the 20-electron stage, and of $BW_{12}O_{40}^{5-}$ to the 18-electron stage, has been followed polarographically (*228*, *229*). As reduction proceeds, successive 'isomeric series' (of unknown structure) appear, such that redox steps linking members of the same series are rapid and reversible, but the passage from one series to another is irreversible. The electrolytic reduction of metatungstate, $H_2W_{12}O_{40}^{6-}$, by up to 36 electrons has been studied (*230*). At all stages, the reduced material can be re-oxidised rapidly and apparently quantitatively to metatungstate. The cationic end-product has been tentatively formulated $H_2W_{12}O_4(OH)_{24}(OH_2)_{12}^{6+}$; if this is correct, it could be isostructural with $Al_{13}O_{40}H_{48}^{7+}$ ($AlAl_{12}O_4(OH)_{24}(OH_2)_{12}^{7+}$) (*58*).

Subsequent remarks refer only to heteropoly blues of a low degree of reduction (up to 6 electrons). Evidence for the retention of the original Keggin or $X_2M_{18}O_{62}^{6-}$ structure comes from the speed and reversibility of the reduction steps (contrasted with the relative kinetic inertness of both oxidised and reduced heteropolyanions, particularly the tungstates, towards isomerisation and degradation)[19]; from the analysis of crystalline salts of the blues[20] (*e.g.* (*231*)); and from X-ray diffraction studies (*232*). Refs. (*46*, *101*, *225*, *226*, *228*–*231* and *233*), which cite earlier work, give the current views concerning polarographic reduction potentials and the numbers of electrons involved, the redox disproportionation of blues, and their *p*H ranges of stability. The following general features may be noted. The maximum charge that a Keggin anion can sustain without decomposition is apparently -8, and that of an $X_2M_{18}O_{62}$ anion is -10 (*46*, *231*). Consequently, the anion will become protonated upon reduction in acidic solution before the limiting charge is attained. Kinetic inertness towards base attack increases as reduction proceeds, causing the effective *p*H range of stability to be extended towards higher *p*H values. Also, as the *p*H is raised some two-electron polarographic waves are observed to undergo separation into two one-electron waves; thus, the conditions for the preparation of several 'odd-electron' blues have been ascertained.

---

19) The A- and B-isomers of $P_2W_{18}O_{62}^{6-}$ give blues which exhibit different ESR spectra, and which are each re-oxidised to the pure original isomer (*97*).

20) A feature of more recent work has been the preparation of pure solutions of the desired blue by controlled electrolysis, either at a cathode potential ascertained from a polarogram or at a higher potential in conjunction with a monitoring electrode; see, for example, Refs. (*97*, *226*, *230*, and *231*).

Within the scope of the present review, however, the heteropoly blues are of greatest interest as mixed-valence complexes; more especially, in view of the numerous 'bronzes' and mixed-valence oxides of molybdenum and tungsten (*223*). In the $XW_{12}O_{40}{}^{(8-n)-}$ series, the variation of the reduction potential for the addition of the first electron with the anion charge can be largely accounted for on electrostatic grounds; the electron enters an orbital whose energy is, to a first approximation, independent of the heteroatom (*46*). The similarity of the optical spectra (resulting from the added electrons) of isoelectronic blues of either the $XMo_{12}O_{40}{}^{(8-n)-}$ or of the $XW_{12}O_{40}{}^{(8-n)-}$ series agrees with this. The question then is: can M(V) and M(VI) atoms be distinguished in the blue anions (that is, remembering that all M atoms are in octahedral sites, are the anions Class II (*223*) mixed-valence complexes?), or do the first and subsequent electrons enter delocalised multicentre orbitals (Class III-A)? The one- and two-electron blues of the $XW_{12}O_{40}{}^{(8-n)-}$ anions (X = Si, P, Co(II), Fe(III), H_2) and $X_2W_{18}O_{62}^{6-}$ anions certainly have Class II character (*97*):

(i) The intensity of absorption in the visible region is roughly proportional to the number of added electrons. The shape and position of the band, which is assigned to W(V) → W(VI) intervalence charge transfer, is little affected. The strong absorption near 38,000 cm^{-1} (oxygen to tungsten(VI) charge transfer) lessens in intensity as electrons are added.

(ii) The sharp proton resonance signal of $H_2W_{12}O_{40}^{6-}$ (*65*) is not seen for the one-electron blue, but reappears, shifted slightly upfield, in the two-electron blue.

(iii) The one-electron blues (X = Si, P, H_2) have μ_{eff} values in the region 1.3—1.4 B.M.; the corresponding two-electron blues are diamagnetic. The ESR g values for the one-electron blues (in the range 1.79—1.90) agree with the presence of W(V) atoms in tetragonal sites. The two-electron blues from $Co^{II}W_{12}O_{40}^{6-}$ and $Fe^{III}W_{12}O_{40}^{5-}$ give the Co(II) and Fe(III) ESR signals of the oxidised anions[21]; the other two-electron blues give no ESR signal. (The two-electron blue $SiMo_{12}O_{40}^{6-}$ gives a weak ESR signal; the g values for this and other 12-molybdosilicate blues agree with the presence of Mo(V) atoms (*234*)). It was concluded that the added electrons are trapped at specific tungsten atoms, but that electron 'hopping' occurs (the lifetime of a particular W(V) atom being about 10^{-8} *s* at 77 K) which accounts for the observed ESR line breadths (100—200 G). The evident electron-pairing in the two-electron blues is by superexchange and *not* by a direct tungsten-tungsten bond. The latter

[21]) The presence of Fe(III) in the tungstoferrate blue had been predicted (*50*).

would require considerable distortion of the Keggin anion, which is not revealed by X-ray studies of the metatungstate blues (*232*).

The intensities of the optical absorption bands of the $P_2Mo_{18}O_{62}^{6-}$ and $P_2W_{18}O_{62}^{6-}$ blues are proportional to the extent of reduction for up to six and four electrons respectively, and the near-UV bands diminish in intensity. This agrees with the Class II mixed-valence description just given. However the absorption of the six-electron blue $H_4P_2W_{18}O_{62}^{8-}$ is unexpectedly weak (*100*). It has been suggested that in this blue we have the beginning of a transition to Class III-A behaviour, the added electrons becoming delocalised when a sufficient number have been added. The analogy has been drawn (*100*) with the bronzes M_xWO_3 and M_xMoO_3; the former becomes metallic at a lower value of x (the fraction of heavy atoms formally in the $+5$ state).

The reduction of a number of anions of the

$X^{+n}Z^{+z}M_{11}O_{40}H_y^{(14-n-y-z)-}$ type has been demonstrated. For example, $SiAlW_{11}O_{40}H_2^{5-}$ (*202*), $SiCo^{II}Mo_{11}O_{40}H_2^{6-}$ (*208*), $SiCr^{III}Mo_{11}O_{40}H_2^{5-}$ (*207*), and $SiCo^{II}W_{11}O_{40}H_2^{6-}$ (*55*)

all exhibit either two or three two-electron polarographic reduction waves, and in the last two cases crystalline salts have been isolated following controlled electrolytic reduction. The $SiW_{11}O_{39}^{8-}$, $PW_{11}O_{39}^{7-}$, and $P_2W_{17}O_{61}^{10-}$ anions (*235*), which are believed to be closely related to the Keggin and 18-series anions (sections 2.3, 2.4), and the $XW_{11}O_{38}H^{6-}$ anions of currently unknown structure (X = Sb(III), Bi(III)) (*236*) are also reducible, although the products slowly transform to reduced 12- or 18-tungstates. A salt $(Me_4N)_6SiMo_2^{V}Mo_8^{VI}Sn_2^{IV}O_{37+x}H_{2x} \cdot nH_2O$ has been isolated (*237*) following the Sn(II) reduction of α—$H_4SiMo_{12}O_{40}$. In none of the above cases have the results of studies bearing on the mixed-valence nature of the reduced anions been published.

6. References

1. *Souchay, P.:* Polyanions et polycations. Paris: Gauthiers-Villars 1963.
2. — Ions minéraux condensés. Paris: Masson 1969.
3. *Kepert, D. L.:* The early transition elements. New York: Academic Press 1972.
4. *Evans, H. T., Jr.:* Perspectives Struct. Chem. *4*, 1 (1971).
5. *Pope, M. T., Dale, B. W.:* Quart. Rev. *22*, 527 (1968).
6. *Tsigdinos, G. A.:* Bulletin Cdb-12a, Climax Molybdenum Co., Greenwich, Conn., 1969.
7. — Bulletin Cdb-14, Climax Molybdenum Co., Greenwich, Conn., 1969.
8. *Keggin, J. F.:* Nature *131*, 908 (1933); Proc. Roy. Soc. (London) A *144*, 75 (1934).
9. *Bradley, A. J., Illingworth, J. W.:* Proc. Roy. Soc. (London) A *157*, 113 (1936).

10. *Yannoni, N. F.:* Thesis, Boston University, 1961.
11. *Eriks, K., Yannoni, N. F., Agarwala, U. C., Simmons, V. E., Baker, L. C. W.:* Acta Cryst. *13*, 1139 (1961).
12. *Smith, P. M.:* Thesis, Georgetown University, 1972; Diss. Abstr. Int. B, *32*, 5136 (1972).
13. *Dawson, B.:* Acta Cryst. *6*, 113 (1953).
14. *Weakley, T. J. R., Evans, H. T., Jr., Showell, J. S., Tourné, G. F., Tourné, C. M.:* J. C. S. Chem. Commun. *1973*, 139.
15. *Strandberg, R.:* Acta Chem. Scand., *27*, 1004 (1973).
16. *Hedman, B.:* Acta Chem. Scand., in the press.
17. *Waugh, J. L. T., Shoemaker, D. P., Pauling, L.:* Acta Cryst. *7*, 438 (1954).
18. *Evans, H. T., Jr.:* J. Am. Chem. Soc. *90*, 3275 (1968).
19. *Hau, H. H.-K.:* Thesis, Boston University, 1970; Diss. Abstr. Int. B, *31*, 2600 (1970).
20. *Agarwala, U. C.:* Thesis, Boston University, 1960.
21. *Perloff, A.:* Inorg. Chem. *9*, 2228 (1970).
22. *Evans, H. T., Jr., Showell, J. S.:* J. Am. Chem. Soc. *91*, 6881 (1969).
23. *Dexter, D. D., Silverton, J. V.:* J. Am. Chem. Soc. *90*, 3589 (1968).
24. *Booth, H. S.* (ed.): Inorganic syntheses, pp. 127, 129, 132. New York: McGraw-Hill 1939.
25. *Brauer, G.:* Handbook of preparative inorganic chemistry, 2nd edit., Vol. 2. New York: Academic Press 1965.
26. *Baker, L. C. W., Loev, B., McCutcheon, T. P.:* J. Am. Chem. Soc. *72*, 2374 (1950).
27. — In: Advances in the chemistry of the coordination compounds, p. 604 (ed. *Kirschner, S.*). New York: MacMillan 1961.
28. *Ganelina, E. S., Nerevyatkina, N. I.:* Zh. Neorgan. Khim. *10*, 483 (1965). — *Ganelina, E. S., Krasnopol'skaya, M. B.:* Zh. Neorgan. Khim. *15*, 1055 (1970). — *Ganelina, E. S., Borgoyakov, V. A.:* Zh. Neorgan. Khim. *16*, 214 (1971).
29. *Krasnopol'skaya, M. B., Ganelina, E. S.:* Uch. Zap. Leningr. Gos. Ped. Inst. *1969*, 119.
30. *Evans, H. T., Jr.:* Inorg. Chem. *5*, 967 (1966). — *Weiss, G.:* Z. Anorg. Allgem. Chem. *368*, 279 (1969). — *Allman, R.:* Acta Cryst. B *27*, 1393 (1971).
31. *Lindquist, I.:* Acta Chem. Scand. *4*, 1066 (1950); *14*, 960 (1960). — *Gatehouse, B. M., Leverett, P.:* J. Chem. Soc. (A) *1971*, 2107.
32. *Anderson, G., Magneli, A.:* Acta Chem. Scand. *4*, 793 (1950).
33. *Barrett, A. S.:* Thesis, Boston University, 1972; Diss. Abstr. Int. B, *33*, 1475 (1972).
34. *Linnett, J. W.:* J. Chem. Soc. *1961*, 3796.
35. *Kepert, D. L.:* Inorg. Chem. *8*, 1556 (1969).
36. *Petrini, G., Pilati, O., Giordano, N.:* Chem. Ind. (Milan) *50*, 1002 (1968). — *Shakova, Z. F., Morosanova, S. A., Zakharova, V. F., Kurenkova, O. N.:* Zh. Neorgan. Khim. *14*, 3053 (1969). — *Polotebnova, N. A., Derkach, L. V.:* Izv. Vyssh. Ucheb. Zaved., Khim. Tekhnol. *12*, 1463 (1969)
37. *Ganelina, E. S.:* Zh. Neorgan. Khim. *7*, 1570 (1962). — *Ganelina, E. S., Reznikova, T. N.:* Zh. Neorgan. Khim. *8*, 1891 (1963). — *Ganelina, E. S., Krasnopol'skaya, M. B.:* Zh. Neorgan. Khim. *14*, 1611 (1969).
38. *Illingworth, J. W., Keggin, J. F.:* J. Chem. Soc. *1935*, 575.
39. *Souchay, P., Tchakirian, A.:* Ann. Chim. *1*, 249 (1946).
40. *Liberti, A., Santoro, A.:* Ric. Sci. *24*, 2079 (1954). — *Liberti, A., Giombini, G., Cervone, E.:* Ric. Sci. *25*, 883 (1955).

41. *Shakova, Z. F., Semenovskaya, E. N., Timofeeva, E. N.:* Zh. Neorgan. Khim. *6*, 330 (1961). — *Shakova, Z. F., Semenovskaya, E. M.:* Zh. Neorgan. Khim. *7*, 1084 (1962); *13*, 1887 (1968).
42. *Brown, D. H., Mair, J. A.:* J. Chem. Soc. *1958*, 2597; *1962*, 1512, 3946. — *Brown, D. H.:* J. Chem. Soc. *1962*, 3281.
43. — J. Chem. Soc. *1962*, 3322.
44. — J. Chem. Soc. *1962*, 4408.
45. *Mair, J. A.:* J. Chem. Soc. *1950*, 2364. — *Mair, J. A., Waugh, J. L. T.:* J. Chem. Soc. *1950*, 2372.
46. *Pope, M. T., Varga, G. M.:* Inorg. Chem. *5*, 1249 (1966).
47. *Wexell, D. R., Pope, M. T.*: Chem. Commun. *1971*, 886.
48. *Souchay, P., Ciabrini, J.-P.:* Compt. Rend. C. *269*, 1525 (1969).
49. *Wyckoff, R. W. G.:* Crystal structures, Vol. 3, p. 887. New York: Interscience 1951.
50. *Simmons, V. E.:* Thesis, Boston University, 1963.
51. *Santos, J. A.:* Proc. Roy. Soc. (London) A *150*, 309 (1935).
52. *Souchay, P.:* Bull. Soc. Chim. France *18*, 365 (1951).
53. — *Tézé, A., Hervé, G.:* Compt. Rend. C *275*, 1013 (1972).
54. *Chalmers, R. A., Sinclair, A. G.:* Anal. Chim. Acta *33*, 384 (1965).
55. *Baker, L. C. W., Figgis, J. S.:* J. Am. Chem. Soc. *92*, 3794 (1970).
56. *Yamamura, K., Sasaki, Y.:* J. C. S. Chem. Commun. 1973, 648.
57. *Jahr, K. F.:* Naturwissenschaften *29*, 505, 528 (1941). — *O'Daniell, H.:* Z. Krist. *104*, 225 (1942).
58. *Johansson, G.:* Acta Chem. Scand. *14*, 771 (1960).
59. *Strickland, J. D. H.:* J. Am. Chem. Soc. *74*, 862, 868, 872 (1952). — *Fruchart, J.-M., Souchay, P.:* Compt. Rend. C *266*, 1571 (1968). — *Biquard, M., Souchay, P.:* Bull. Soc. Chim France *1971*, 437.
60. *Massart, R., Souchay, P.:* Compt. Rend. C *257*, 1297 (1963). — *Contant, R.:* Compt. Rend. C *267*, 1479 (1968).
61. *Souchay, P., Contant, R.:* Compt. Rend. C *265*, 723 (1967). — *Halasz, A., Pungor, E.:* Talanta *18*, 569 (1971).
62. *Signer, R., Gross, H.:* Helv. Chim Acta *17*, 1076 (1934).
63. *Schott, G., Harzdorf, C.:* Z. Anorg. Allgem. Chem. *288*, 15 (1956). — *Mooney, R. W., Chiola, V., Hoffmann, C. W. W., Vanderpool, C. D.:* J. Electrochem. Soc. *109*, 1179 (1962). — *Rode, E. Y., Ivanova, M. M., Karpov, V. N.:* Dokl. Akad. Nauk. SSSR *176*, 848 (1967).
64. *Spitsyn, V. I., Chuvaev, V. F., Kabanov, V. Y.:* Dokl. Akad. Nauk. SSSR *152*, 153 (1963). — *Glemser, O., Holznagel, W., Holtje, W., Schwarzmann, E.:* Z. Naturforsch. b *20*, 725 (1965). — *Spitsyn, V. I., Lunk, H.-J., Chuvaev, V. F., Kolli, I. D.:* Z. Anorg. Allgem. Chem. *370*, 191 (1969).
65. *Pope, M. T., Varga, G. M.:* Chem. Commun. *1966*, 653.
66. *Kokorin, A. I., Polotebnova, N. A.:* Zh. Obshch. Khim. *26*, 3 (1956).
67. *Ablov, A. V., Malinovskii, T. I., Dedyu, V. I.:* Zh. Neorgan. Khim. *4*, 367 (1959).
68. *Kokorin, A. I.:* Zh. Obshch. Khim. *24*, 966 (1954). — *Kokorin, A. I., Polotebnova, N. A.:* Zh. Obshch. Khim. *27*, 304 (1957). — *Courtin, P., Chauveau, F., Souchay, P.:* Compt. Rend. C *258*, 1247 (1964). — *Souchay, P., Chauveau, F., Courtin, P.:* Bull. Soc. Chim. France *1968*, 2384.
69. *Tsigdinos, G. A., Hallada, C. J.:* Inorg. Chem. *7*, 437 (1968).
70. *Courtin, P.:* Bull. Soc. Chim. France *1968*, 4799; Rev. Chim. Miner. *8*, 75 (1971).
71. *Begalieva, D. U., Bekturov, A. B., Il'yasova, A. K.:* Zh. Neorgan. Khim. *16*, (1971).

72. *Smith, D. P., Pope, M. T.:* Inorg. Chem. *12*, 331 (1973).
73. *Suteu, A., Mirza, L.:* Rev. Roumaine Chim. *15*, 1563 (1970).
74. *Kokorin, A. I., Dimant, R. L.:* Zh. Obshch. Khim. *24*, 971 (1954). — *Kokorin, A. I., Polotebnova, N. A.:* Zh. Obshch. Khim. *24*, 1137, 1718 (1954). — *Motorkina, R. K.:* Zh. Neorgan. Khim. *2*, 92 (1957). — *Polotebnova, N. A.:* Zh. Neorgan. Khim. *10*, 1498 (1965). — *Radul, K. K., Polotebnova, N. A., Bardin, M. B.:* Zh. Vses. Khim. Obshch. *14*, 585 (1969).
75. *Suteu, A., Cristea, G.:* Rev. Roumaine Chim. *14*, 209 (1969).
76. *Flynn, C. M., Pope, M. T.:* Inorg. Chem. *10*, 2745 (1971); *11*, 1950 (1972).
77. *Baker, L. C. W., Baker, V. S., Eriks, K., Pope, M. T., Shibata, M., Rollins, O. W., Fang, J. H., Koh, L. L.:* J. Am. Chem. Soc. *88*, 2329 (1966).
78. *Weakley, T. J. R., Malik, S. A.:* J. Inorg. Nucl. Chem. *29*, 2935 (1967).
79. *Ripan, R., Puscasiu, M.:* Z. Anorg. Allgem. Chem. *358*, 82 (1968).
80. *Tourné, C. M., Tourné, G. F., Malik, S. A., Weakley, T. J. R.:* J. Inorg. Nucl. Chem. *32*, 3875 (1970).
81. — — Compt. Rend. C *266*, 1383 (1968).
82. *Rollins, O. W.:* Thesis, Georgetown University, 1966; Diss. Abstr. Int. B, *26*, 7024 (1966).
83. — J. Inorg. Nucl. Chem. *33*, 75 (1971).
84. *Tourné, C. M.:* Compt. Rend. C *266*, 702 (1968).
85. — *Tourné, G. F.:* Bull. Soc. Chim. France *1969*, 1124.
86. *Souchay, P.:* Ann. Chim. *20*, 73 (1945); *1*, 232 (1946).
87. — Ann. Chim. *2*, 203 (1947).
88. — *Faucherre, J.:* Bull. Soc. Chim. France *18*, 355 (1951).
89. *Biquard, M., Lamache, M.:* Bull. Soc. Chim. France *1971*, 32.
90. *Babko, A. K., Shkaravskii, Y. F.:* Zh. Neorgan. Khim. *7*, 1565 (1962). — *Shkaravskii, Y. F.:* Zh. Neorgan. Khim. *8*, 2668 (1963); Ukr. Khim. Zh. *30*, 241 (1964). — *Reznik, B. E., Vorotyagina, V. D.:* Zh. Neorgan. Khim. *13*, 3080 (1968). — *Barkovskii, V. F., Skornyakova, O. N.:* Zh. Neorgan. Khim. *14*, 1046 (1969). — *Barkovskii, V. F., Velikanova, T. V.:* Zh. Neorgan. Khim. *15*, 749, 1618 (1970).
91. *Babko, A. K., Shkaravskii, Y. F.:* Zh. Neorgan. Khim. *6*, 2091 (1961). — *Barkovskii, V. F., Zaboeva, M. I.:* Zh. Neorgan. Khim. *10*, 900 (1965). — *Shkaravskii, Y. F.:* Zh. Neorgan. Khim. *10*, 1179 (1965); *11*, 120 (1966).
92. *Shkaravskii, Y. F.:* Zh. Neorgan. Khim. *11*, 797 (1966). — *Barkovskii, V. F., Radovskaya, T. L.:* Zh. Neorgan. Khim. *12*, 991 (1967).
93. *Barkovskii, V. F., Velikanova, T. V.:* Zh. Neorgan. Khim. *11*, 2296 (1966).
94. *Tsigdinos, G. A.:* unpublished work, quoted in Ref. 6.
95. *Wu, H.:* J. Biol. Chem. *43*, 189 (1920).
96. *Langs, D. A., Silverton, J. V.:* unpublished work, quoted in Ref. 96.
97. *Varga, G. A., Papaconstantinou, E., Pope, M. T.:* Inorg. Chem. *9*, 662 (1970).
98. *Nikitina, E. A., Tsvetkov, N. A.:* Zh. Neorgan. Khim. *4*, 839 (1959).
99. — — Zh. Neorgan. Khim. *7*, 325 (1962).
100. *Papaconstantinou, E., Pope, M. T.:* Inorg. Chem. *9*, 667 (1970).
101. — — Inorg. Chem. *6*, 1147 (1967).
102. *Souchay, P., Contant, R., Fruchart, J. M.:* Compt. Rend. C *264*, 976 (1967).
103. *Brown, D. H.:* J. Inorg. Nucl. Chem. *14*, 129 (1960).
104. *Rosenheim, A., Wolff, A.:* Z. Anorg. Allgem. Chem. *193*, 64 (1930).
105. *Kehrmann, F., Bohm, E.:* Z. Anorg. Chem. *6*, 386 (1894). — *Kehrmann, F.:* Z. Anorg. Chem. *7*, 422 (1894); *22*, 285 (1899).
106. *Malik, S. A., Weakley, T. J. R.:* J. Chem. Soc. (A) *1968*, 2647.

107. *Pettersson, L.:* Acta Chem. Scand. *25*, 1959 (1971). — *Pettersson, L., Andersson, I., Lyhamn, L., Ingri, N.:* Trans. Roy. Inst. Tech. Stockholm *1972*, 109.
108. *Prasad, S., Swarup, R.:* J. Inorg. Nucl. Chem. *30*, 1337 (1968).
109. *Rosenheim, A., Schapiro, M.:* Z. Anorg. Chem. *129*, 196 (1923).
110. *Michelon, M., Hervé, G.:* Compt. Rend. C *274*, 209 (1972).
111. *Lautié, R.:* Bull. Soc. Chim. France *1*, 105 (1934).
112. *Souchay, P., Cadiot, M., Volfovsky, C.:* Compt. Rend. C *269*, 826 (1969).
113. *Viossat, B., Volfovsky, C., Cadiot, M.:* Compt. Rend. C *273*, 1637 (1971).
114. *Tsigdinos, G. A.:* Thesis, Boston University, 1961.
115. *Wood, J. W.:* Thesis, University of East Anglia, 1968.
116. *Anderson, J. S.:* Nature *140*, 850 (1937).
117. *Evans, H. T.,Jr.:* J. Am. Chem. Soc. *70*, 1291 (1948).
118. *La Ginestra, A., Giannetta, F., Fiorucci, P.:* Gazz. Chim. Ital. *98*, 1197 (1968).
119. *Wolfe, C. W., Block, M. L., Baker, L. C. W.:* J. Am. Chem. Soc. *77*, 2200 (1955).
120. *Rollins, O. W., Earley, J. E.:* J. Am. Chem. Soc. *81*, 5571 (1959).
121. *Rosenheim, A., Pieck, M., Pinsker, J.:* Z. Anorg. Chem. *96*, 139 (1916).
122. *Shimura, Y., Ito, H., Tsichida, R.:* J. Chem. Soc. Japan *75*, 560 (1954).
123. *Matijevic, E., Kerker, M., Beyer, H., Theubert, F.:* Inorg. Chem. *2*, 581 (1963).
124. *Ouchi, A., Matsumoto, H., Yoshino, Y.:* J. Chem. Soc. Japan *83*, 438 (1962).
125. *Baker, L. C. W., Shibata, M.:* Abstracts, 100th Amer. Chem. Soc. meeting, New York, 1960.
126. *Prasad, S., Garg, V. N.:* Indian J. Chem. *5*, 575 (1967). — *Shakova, Z. F., Zakharova, V. F., Morosanova, S. A.:* Zh. Neorgan. Khim. *14*, 1895 (1969).
127. *Shakova, Z. F., Ku T'han Long:* Zh. Neorgan. Khim. *9*, 1848 (1964).
128. *Volfovsky, C., Cadiot, M.:* Compt. Rend. C *274*, 62 (1972).
129. *Ama, T., Hidaka, J., Shimura, Y.:* Bull. Soc. Chem. Japan *43*, 2654 (1970).
130. *Barbieri, G. A.:* Atti Accad. Lincei *22*, 781 (1913). — *Shakova, Z. F., Gavrilova, S. A., Zakharova, V. A.:* Zh. Neorgan. Khim. *7*, 1752 (1962). — *Torchenkova, E.A., Baidala, P., Smurova, V. S., Spitsyn, V. I.:* Dokl. Akad. Nauk. SSSR *199*, 120 (1971).
131. *Baidala, P., Smurova, V. S., Torchenkova, E. A., Spitsyn, V. I.:* Dokl. Akad. Nauk. SSSR *197*, 339 (1971).
132. *Barbier, A., Malaprade, L.:* Compt. Rend. C *256*, 168 (1963).
133. *McKean, L., Pope, M. T.:* Inorg. Chem., in the press.
134. *Nguyen-Dieu, Torchenkova, E. A., Spitsyn, V. I.:* Dokl. Akad. Nauk. SSSR *198*, 1350 (1971).
135. *Torchenkova, E. A., Stepanova, G. G., Spitsyn, V. I.:* Dokl. Akad. Nauk. SSSR *157*, 1167 (1964); Rec. Chem. Progr. *31*, 89 (1970).
136. *Dubois, S., Souchay, P.:* Ann. Chim. *2*, 105 (1948).
137. *Burnel, D., Lienard, J., Malaprade, L., Voisin, C.:* Bull. Soc. Chim. France *1969*, 412.
138. *Brown, D. H., Hyde, A. J.:* J. Chem. Soc. *1962*, 3186.
139. *Sillen, L. G.:* In: Coordination chemistry (ed. Martell, A. E.), Vol. 1, p. 491. New York: Van Nostrand Reinhold 1971.
140. *Flynn,* C. M., *Pope, M. T.:* Inorg. Chem. *10*, 2524 (1971).
141. *Ripan, R., Todorut, I.:* Roczniki Chem. *38*, 1587 (1964); Rev. Roumaine Chem. *11*, 691, 1279 (1966). — *Ripan, R., Botar, A.:* Rev. Roumaine Chem. *15*, 1529, 1537 (1970). — *Marcu, G., Botar, A.:* Rev. Roumaine Chem. *16*, 1341 (1971).
142. *Peacock, R. D., Weakley, T. J. R.:* J. Chem. Soc. (A) *1971*, 1836.
143. *Botar, A.:* personal communication, 1972.
144. *Souchay, P., Leray, M., Hervé, G.:* Compt. Rend. C *271*, 1337 (1970).

145. *Kehrmann, F., Freinkel, M.:* Ber. *24*, 2326 (1891). — *Kehrmann, F.:* Z. Anorg. Chem. *1*, 423 (1892); *22*, 285 (1899).
146. *Gmelins:* Handbuch der Anorganischen Chemie, 8th edit., parts 53 (1935) and 54 (1933). Berlin: Verlag Chemie.
147. *Mellor, J. W.:* Comprehensive treatise on inorganic and theoretical chemistry, vol. 11. London: Longman, Green 1931.
148. *Wada, T.:* Compt. Rend. *259*, 553 (1964).
149. *Chuvaev, V. F., Kabanov, V. Y., Spitsyn, V. I.:* Dokl. Akad. Nauk. SSSR *154*, 908 (1964).
150. *Wada, T.:* Compt. Rend. B *263*, 51 (1966).
151. *West, S. F., Audrieth, L. F.:* J. Phys. Chem. *59*, 1069 (1955). — *Babad-Zakhryapin, A. A.:* Zh. Neorgan. Khim. *3*, 2313 (1958).
152. *Krotov, N. A., Rode, E. Y.:* Zh. Neorgan. Khim. *8*, 1722 (1963).
153. *Ivanova, M. M., Rode, E. Y.:* Zh. Neorgan. Khim. *11*, 223 (1966). — *Ivanova, M. M., Rode, E. Y., Balagina, G. M.:* Zh. Neorgan. Khim. *11*, 320 (1966). — *Rode, E. Y., Ivanova, M. M., Karpov, V. N.:* Dokl. Akad. Nauk. SSSR *176*, 848 (1967).
154. *Ripan, R., Puscasiu, M., Stanescu, D., Boian, P.:* Z. Anorg. Allgem. Chem. *391*, 183 (1972).
155. *Marsmann, H., Hahn, H.:* Z. Naturforsch. *21*, 188 (1966).
156. *Puscasiu, M.:* Dokl. Akad. Nauk SSSR *182*, 634 (1968).
157. *Stanescu, D.:* Z. Anorg. Allgem. Chem. *366*, 104 (1969).
158. *La Ginestra, A., Cerri, R.:* Gazz. Chim. Ital. *95*, 26 (1965). — *La Ginestra, A., Cerri, R., Gianetta, F., Fiorucci, P.:* J. Thermal Anal. *2*, 107 (1970).
159. *Baker, L. C. W., Foster, G., Tan, W., Scholnick, F., McCutcheon, T. P.:* J. Am. Chem. Soc. *77*, 2136 (1955).
160. *Chuvaev, V. F., Spitsyn, V. I.:* Dokl. Akad. Nauk. SSSR *166*, 160 (1966).
161. — *Lunk, K. I., Spitsyn, V. I.:* Dokl. Akad. Nauk. SSSR *181*, 1156 (1968).
162. *Spitsyn, V. I., Bakhchisaraitseva, S. A.:* Zh. Fiz. Khim. *40*, 1961 (1966).
163. *Chuvaev, V. F., Baidala, P., Torchenkova, E. A., Spitsyn, V. I.:* Dokl. Akad. Nauk SSSR *196*, 1097 (1971).
164. *Baker, L. C. W., Gallagher, G. A., McCutcheon, T. P.:* J. Am. Chem. Soc. *75*, 2493 (1953).
165. *Spitsyn, V. I., Chuvaev, V. F., Bakhchisaraitseva, S. A.:* Dokl. Akad. Nauk SSSR *160*, 658; *165*, 1126 (1965).
166. *Chuvaev, V. F., Shinnik, G. N., Polotebnova, N. A., Spitsyn, V. I.:* Dokl. Akad. Nauk. SSSR *204*, 1403 (1972).
167. *Nikitina, E. A., Kulanova, N. E.:* Zh. Neorgan. Khim. *4*, 564, 571 (1959). — *Ripan, R., Liteanu, C., Popescu, R.:* Zh. Neorgan. Khim. *4*, 1300 (1959). — *Ripan, R., Liteanu, C., Neagu, V., Popescu, R.:* Zh. Neorgan. Khim. *6*, 246 (1961). — *Nikitina, E. A., Tsvetkov, N. A.:* Zh. Neorgan. Khim. *10*, 2648 (1965). — *Nikitina, E. A., Sesiashvili, K. M.:* Zh. Neorgan. Khim. *14*, 1885 (1969).
168. *Spitsyn, V. I.:* Z. Anorg. Allgem. Chem. *304*, 196 (1960). — *Rubel, M. P., Spitsyn, V. I.:* Zh. Neorgan. Khim. *5*, 2770, 2774 (1960). — *Spitsyn, V. I., Kosmodem'yanskaya, G. V.:* Zh. Neorgan. Khim. *10*, 657 (1965).
169. *Rode, E. Y., Krotov, N. A.:* Zh. Neorgan. Khim. *8*, 939 (1963). — *Rode, E. Y., Sokolova, M. P.:* Zh. Neorgan. Khim. *8*, 1883 (1963). — *Ivanova, M. M., Rode, E. Y.:* Zh. Neorgan. Khim. *11*, 223 (1966).
170. *Sharpless, N. E., Munday, J. S.:* Anal. Chem. *29*, 1619 (1957).
171. *Brown, D. H.:* Spectrochim. Acta *19*, 585 (1963).
172. *Bakhchisaraitseva, S. A., Kabanov, V. Y., Spitsyn, V. I.:* Dokl. Akad. Nauk. SSSR *174*, 624 (1967).

173. *Lange, G., Hahn, H., Dehnicke, K.:* Z. Naturforsch. *24*b, 1498 (1969).
174. *Rabette, P., Olivier, D.:* Rev. Chim. Miner. *7*, 181 (1970).
175. *Herzberg, G.:* Infra-red and raman spectra of polyatomic molecules, pp. 194–199. New York: Van Nostrand 1945.
176. *Weakley, T. J. R.:* unpublished work.
177. *Yukhnevich, G. V.:* Zh. Neorgan. Khim. *6*, 231 (1961).
178. *Baker, L. C. W., Simmons, V. E.:* J. Am. Chem. Soc. *81*, 4744 (1959).
179. *Wood, D. L., Remeika, J. P.:* J. Chem. Phys. *46*, 3595 (1967).
180. *Ray, P., Bhaduri, A., Sarma, B.:* J. Indian Chem. Soc. *25*, 51 (1948).
181. *So, H., Pope, M. T.:* Inorg. Chem. *11*, 1441 (1972).
182. *Peacock, R. D., Weakley, T. J. R.:* J. Chem. Soc. (A) *1971*, 1937.
183. *Baker, L. C. W., Weakley, T. J. R.:* J. Inorg. Nucl. Chem. *28*, 447 (1966).
184. *Shimura, Y., Ito, H., Tsuchida, R.:* J. Chem. Soc. Japan *75*, 560 (1954).
185. *Brown, D. H.:* Spectrochim. Acta *19*, 1683 (1963).
186. *Shimura, Y., Tsuchida, R.:* Bull. Chem. Soc. Japan *30*, 502 (1957).
187. *Jørgensen, C. K.:* Absorption spectra and chemical bonding in complexes, pp. 113, 138. Oxford: Pergamon 1962.
188. Ref. 187. chapter 15.
189. *Ferguson, J., Knox, K., Wood, D. L.:* J. Chem. Phys. *35*, 2236 (1961).
190. *Wood, D. L., Imbusch, G. F., Macfarlane, R. M., Kisliuk, P., Larkin, D. M.:* J. Chem. Phys. *48*, 5255 (1968).
191. *Pappalardo, R., Wood, D. L., Linares, R. C.:* J. Chem. Phys. *35*, 2041 (1961).
192. *Ferguson, J., Wood, D. L., Van Uitert, L. G.:* J. Chem. Phys. *51*, 2904 (1969).
193. *Peacock, R. D.:* Chem. Phys. Letters *7*, 187 (1970), *10*, 134 (1971); J. Chem. Soc. (A) *1971*, 2028; J. C. S. Faraday II *68*, 169 (1972); Mol. Phys. *25*, 817 (1973).
194. *Paulusz, A. G., Burrus, H. L.:* Chem. Phys. Letters *17*, 527 (1972).
195. *Meredith, P. L., Palmer, R. A.:* Chem. Commun. *1969*, 1337. — *Grinter, R., Harding, M. J., Mason, S. F.:* J. Chem. Soc. (A) *1970*, 667.
196. *Rasmussen, P. G., Beem, K. M., Hornyak, E. H.:* J. Chem. Phys. *50*, 3647 (1969).
197. *Byfleet, C. R., Lin, W. C., McDowell, C. A.:* Mol. Phys. *18*, 363 (1970).
198. *Spijkerman, J. J.:* Symposia Faraday Soc. *1*, 134 (1967).
199. *So, H., Pope, M. T.:* J. Chem. Phys. *55*, 2786 (1971).
200. *McGarvey, B. R.:* Transition Metal Chem. *3*, 119, 169 (1966).
201. *Baker, L. C. W., McCutcheon, T. P.:* J. Am. Chem. Soc. *78*, 4503 (1956).
202. *Tézé, A., Souchay, P.:* Rev. Chim. Miner. *7*, 539 (1970).
203. *Bauchet, M., Tourné, C. M., Tourné, G. F.,:* Compt. Rend. C *275*, 407 (1972).
204. *Smith, D. P., So, H., Bender, J., Pope, M. T.:* Inorg. Chem. *12*, 685 (1973).
205. *Souchay, P., Bertho, G.:* Compt. Rend. C *262*, 42 (1966).
206. *Colom, R.:* Thesis, Université Sci. Tech. Languedoc, Montpellier, 1972.
207. *Fournier, M., Massart, R.:* Compt. Rend. C *276*, 1517 (1973).
208. *Leyrie, M., Fournier, M., Massart, R.:* Compt. Rend. C *273*, 1569 (1971).
209. *Rigny, P., Pinsky, L., Weulersse, J.-M.:* Compt. Rend. C *276*, 1223 (1973).
210. *Souchay, P., Lamache, M., Petit, M.:* Bull. Soc. Chim. France *1971*, 37.
211. *Petit, M., Massart, R.:* Compt. Rend. C *268*, 1860 (1969).
212. *Rollins, O. W.:* Inorg. Chem. *11*, 3114 (1972).
213. *Weakley, T. J. R.:* J. C. S. Dalton *1973*, 341.
214. *Ripan, R., Duca, A., Stanescu, D., Puscasiu, M.:* Z. Anorg. Allgem. Chem. *347*, 333 (1966). — *Ripan, R., Stanescu, D.:* Rev. Chim. Miner. *4*, 899 (1967).
215. — *Stanescu, D.:* Z. Anorg. Allgem. Chem. *380*, 107 (1971).

216. — *Puscasiu, M., Stanescu, D., Boian, P.:* Z. Anorg. Allgem. Chem. *384*, 297 (1971).
217. — *Stanescu, D., Puscasiu, M.:* Z. Anorg. Allgem. Chem. *391*, 187 (1972).
218. *Malik, S. A.:* J. Inorg. Nucl. Chem. *32*, 2425 (1970).
219. *Charreton, B., Meunier, R.:* Compt. Rend. C *275*, 945 (1972).
220. *Baker, L. C. W., Baker, V. E. S., Wasfi, S. H., Candela, G. A., Kahn, A. H.:* J. Chem. Phys. *56*, 4917 (1972); J. Am. Chem. Soc. *94*, 5499 (1972). — *Wasfi, S. H.:* Thesis, Georgetown University, 1971.
221. *Ripan, R., Puscasiu, M.:* Z. Anorg. Allgem. Chem. *380*, 102 (1971).
222. *Bauchet, M.:* Thesis, Université Sci. Tech. Languedoc, Montpellier, 1971.
223. *Robin, M. B., Day, P.:* Advan. Inorg. Chem Radiochem *10*, 247 (1967).
224. *Souchay, P., Massart, R.:* Compt. Rend. C *253*, 1699 (1961). — *Stonehart, P., Koren, J. G., Brinen, J. S.:* Anal. Chim. Acta *40*, 65 (1968). — *Biquard, M., Souchay, P.:* Bull. Soc. Chim. France *1971*, 437.
225. *Massart, R.:* Ann. Chim. *4*, 285 (1969).
226. *Hervé, G.:* Ann. Chim. *6*, 219 (1971).
227. *Pope, M. T.:* Inorg. Chem. *11*, 1973 (1972).
228. *Fruchart, J.-M., Hervé, G.:* Ann. Chim. *6*, 337 (1971).
229. *Hervé, G.:* Ann. Chim. *6*, 287 (1971).
230. *Tourné, C. M.:* Bull. Soc. Chim. France *1967*, 3196, 3199, 3214.
231. *Papaconstantinou, E., Pope, M. T.:* Inorg. Chem. *6*, 1152 (1967).
232. *Tourné, C. M., Souchay, P.:* Compt. Rend. C *263*, 1142 (1966).
233. *Massart, R.:* Ann. Chim. *3*, 507 (1968). — *Massart, R., Hervé, G.:* Chim. Miner. *5*, 501 (1968). — *Launay, J. P., Souchay, P., Boyer, M.:* Collection Czech. Chem. Commun. *36*, 740 (1971).
234. *Rabette, P., Ropars, C., Grivet, J.-P.:* Compt. Rend. C *265*, 153 (1967).
235. *Souchay, P., Tézé, A.:* Compt. Rend. C *268*, 804 (1969).
236. *Michelon, M., Souchay, P., Massart, R., Hervé, G.:* Compt. Rend. C *273*, 1117 (1971).
237. *Fournier, M., Massart, R., Souchay, P.:* Compt. Rend. C *272*, 451 (1971).

Received August 27, 1973

Hydrogen Bonding in Solids.
Correlation of Spectroscopic and Crystallographic Data*

A. Novak

Laboratoire de Chimie Physique, C.N.R.S., 94320, Thiais, France

Table of Contents

I. Introduction

When an AH acid molecule is engaged in an A—H..B hydrogen bond the basic group represents a supplementary attraction potential for the proton and can create a second (II) potential well, as illustrated dia-

* Plenary lecture presented at the First European Crystallographic Meeting, Bordeaux 5—8th September, 1973.

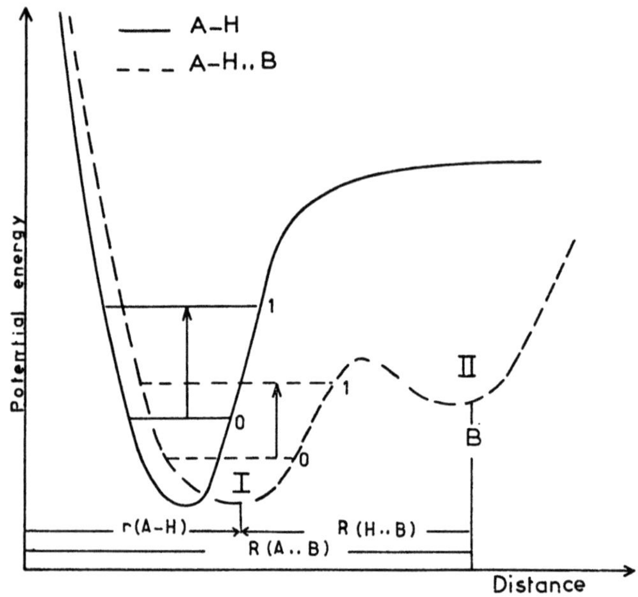

Fig. 1. Qualitative potential curves for the free (———) and hydrogen bonded (— — — —) AH group

grammatically in Fig. 1. The potential energy function of the free AH group is thus modified by the potential B; it becomes broader and the vibrational levels become closer as reflected by the shift of the AH stretching band toward lower frequencies shown by infrared and Raman spectroscopy. At the same time, the proton of the AH group shifts toward B and the equilibrium r (A—H) distance increases while the intermolecular R(A..B) distance decreases to a value less than the sum of the Van der Waals radii of the A and B atoms. The geometrical changes have been detected by neutron and X-ray diffraction methods (*1—4*).

1. Classification of Hydrogen Bonds

We have thus, in addition to the enthalpy of hydrogen bond formation (ΔH), two main criteria for estimating the strength of this interaction: 1. AH stretching frequency ν or its relative shift $(\nu_0-\nu)/\nu_0$, where ν_0 corresponds to the frequency of the "free" AH group and, 2. r(A—H), R(A..B) and R(H..B) distances. It seems convenient to distinguish — by using these criteria — weak (long) and strong (short) hydrogen bonds since their properties are rather different. This distinction appears to be justified also by the fact that the theory of weak hydrogen bonds is more developed than that of strong interactions (*2*).

Table 1 gives an approximate classification of the O—H. . O hydrogen bond system, the system most studied. Here we can distinguish weak, strong, and intermediate hydrogen bonds.

Table 1. Approximate classification of O—H. . . O hydrogen bonds

Hydrogen bond	νOH (cm^{-1}) frequency	Relative [a] shift (%)	RO. . . O(A°) distance	Enthalpy ΔH (kcal/mole)	Examples
Weak	>3200	<12%	2.70	<5	H_2O (ices, hydrates) R—OH (alcohols, phenols)
Intermediate	3100—2800	12—22%	2.70—2.60	6—8	RCOOH (acids)
Strong	2700—700	25—80%	2.60—2.40	>8	$MH(RCOO)_2$ (acid salts)

[a] $(\nu_0-\nu)/\nu_0$ where ν_0 corresponds to the "free" OH stretching frequency.

The type of hydrogen bond may also be classified according to the shape of the potential curve of the proton of an A—H. . B system or to the position of the proton (*2,4*). An acid-base reaction can be represented by a series of equilibria: $AH + B \rightleftarrows A{-}H..B \rightleftarrows (A..H..B) \rightleftarrows A^-..HB^+ \rightleftarrows A^- + HB^+$.

The formation of a hydrogen bond is followed by a transition state of the proton transfer which can give a complex between the BH^+ acid and its conjugated base A^- and finally "free" ions. The AH. . B system containing a weak or moderately strong hydrogen bond has a potential curve similar to that of the Fig. 1. For the $A^-..HB^+$ system the potential curve is similar but with well II deeper than well I. Finally, the potential curve may be symmetric. In all three cases the potential barrier separating the two minima is high and the proton is always closer to one atom than to the other (A or B).

However, in the (A. . H. . B) system (more particularly in those represented by $(A{-}H{-}A)^-$ and $(B{-}H{-}B)^+$) containing strong hydrogen bonds (Table 1), other types of potential curves may exist. This is the case of the "hesitating proton" characterized by a small or zero potential barrier. We can distinguish, 1. the asymmetric double minimum, 2. the symmetric double minimum, and 3. the symmetric single minimum by spectroscopic means (see Section III).

2. Infrared Spectra

The spectroscopic behavior of an O—H..O hydrogen bond which becomes progressively stronger is shown in Fig. 2. The "free" OH stretching frequency of the acetic acid molecule in gaseous state at 430 °K is observed at 3583 cm^{-1} (*5*). It decreases to 2875 cm^{-1} in the acetic acid crystal near

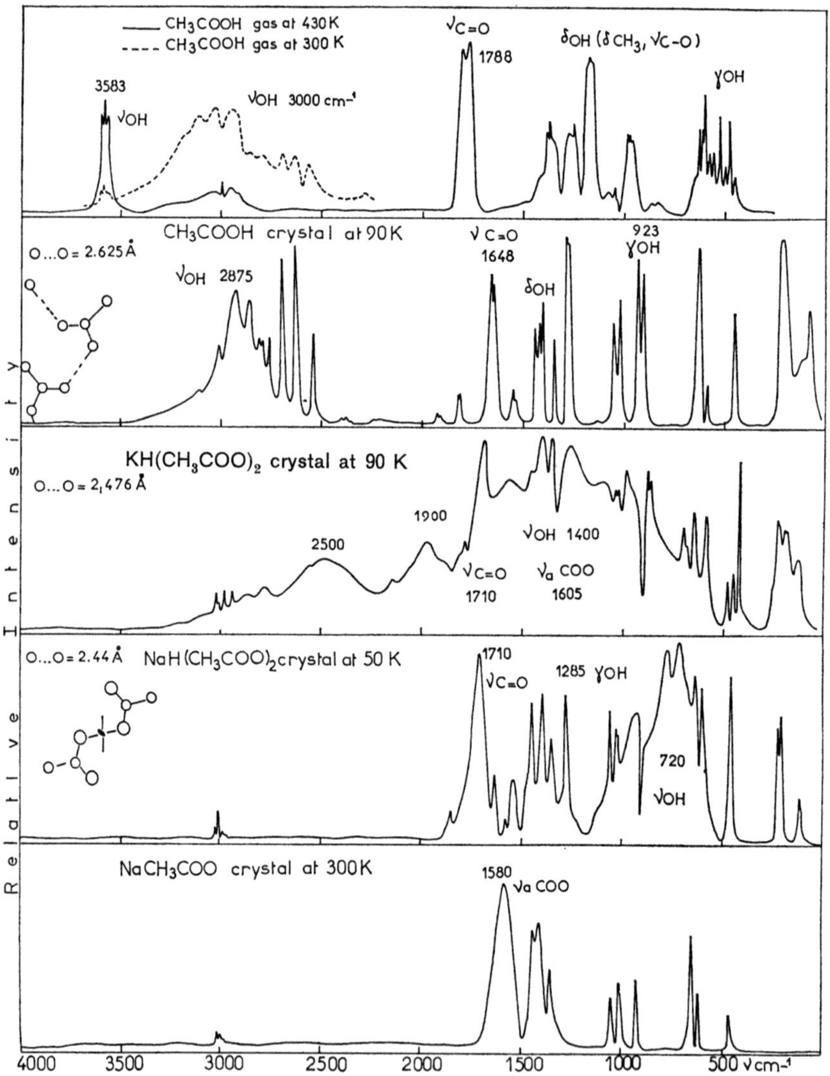

Fig. 2. Infrared spectra of acetic acid and of some of its acid and neutral salts

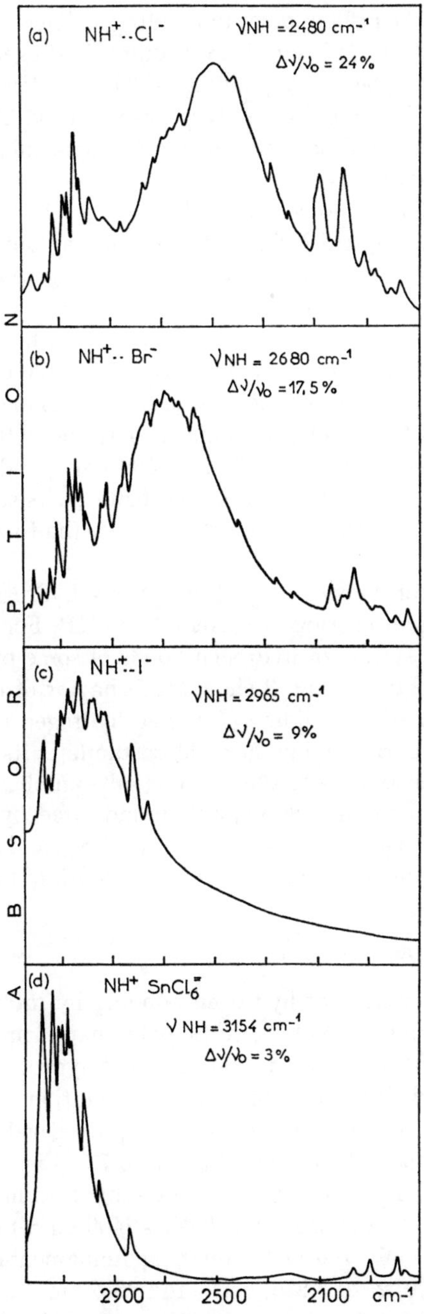

Fig. 3. Infrared spectra of some pyrimidinium salts, $C_4N_2H_5^+X^-$, (a) chloride, (b) bromide, (c) iodide, (d) hexachlorostanate

90 °K (*6*), to 1400 cm^{-1} in potassium hydrogen diacetate, $KH(CH_3COO)_2$ (*7*), and finally to 720 cm^{-1} in sodium hydrogen diacetate, $NaH(CH_3COO)_2$ (*8*). The corresponding R(O..O) distances are 2.63 Å (*9,10*), 2.49 Å (*11*), and 2.44 Å (*12*), respectively. The changes in the intensity, breadth, and structure of the OH stretching absorption band are no less spectacular.

The CH_3COOH monomer molecule shows a νOH band about 30 cm^{-1} wide (P—R distance) which differs little from any other absorption band arising from the same symmetry species. The gaseous dimer at room temperature, however, yields a νOH band whose half-width is almost 400 cm^{-1}, while in the crystal at 90 °K, which contains infinite chains of CH_3COOH molecules, this half-width is "only" 120 cm^{-1}. The complex structure of the band can be ascribed in both cases to combination bands and overtones (*5,6*). The OH stretching absorption in the $KH(CH_3COO)_2$ crystal is observed between 3100 and 500 cm^{-1} and the half-width is estimated to be about 1000 cm^{-1}. The band in its sodium analogue at somehow lower temperature is much narrower (half-width $= \nu^1/_2 \simeq 250$ cm^{-1}). (Fig. 2).

The distinction between strong and weak hydrogen bonds can also be made in the proton transfer system $A^-..^+HB$. For example, consider the $N—H^+..X^-$ system of hydrogen bonds of some pyrimidinium salts. The infrared spectra of Fig. 3 show the behavior of the NH stretching band of protonated pyrimidine (*13*). The hydrogen bond is very weak in the crystal of pyrimidinium hexachlorostannate; its strength increases upon going to the iodide and the bromide salts and becomes considerable in the pyrimidinium chloride crystal, as indicated by the relative shift of the NH stretching frequency of 3%, 9%, 17,5% and 24%, respectively, as well as by the increasing intensity and breadth of the νNH band.

3. Raman Spectra

There are many studies of hydrogen bonding by infrared but very few by Raman techniques. Yet we anticipate that Raman scattering will play an important role in the investigation of hydrogen bonded systems. Some recent work (*14,15*) confirms this expectation.

In particular Raman spectroscopy has proved valuable in the study of the profiles of the AH stretching bands (*16,17*). Fig. 4 shows the Raman spectra in the OH stretching region of some intermediate and strong O—H..O bonds. In the spectrum (3600—2400 cm^{-1}) of the CD_3COOH crystal at 50 °K only the OH stretching fundamental is expected yet five bands at 3030, 2920, 2825, 2662, and 2557 cm^{-1} are observed. They are interpreted, as are their infrared analogues, in terms of Fermi resonances between the OH stretching fundamental and $2\,\delta$OH, δOH +

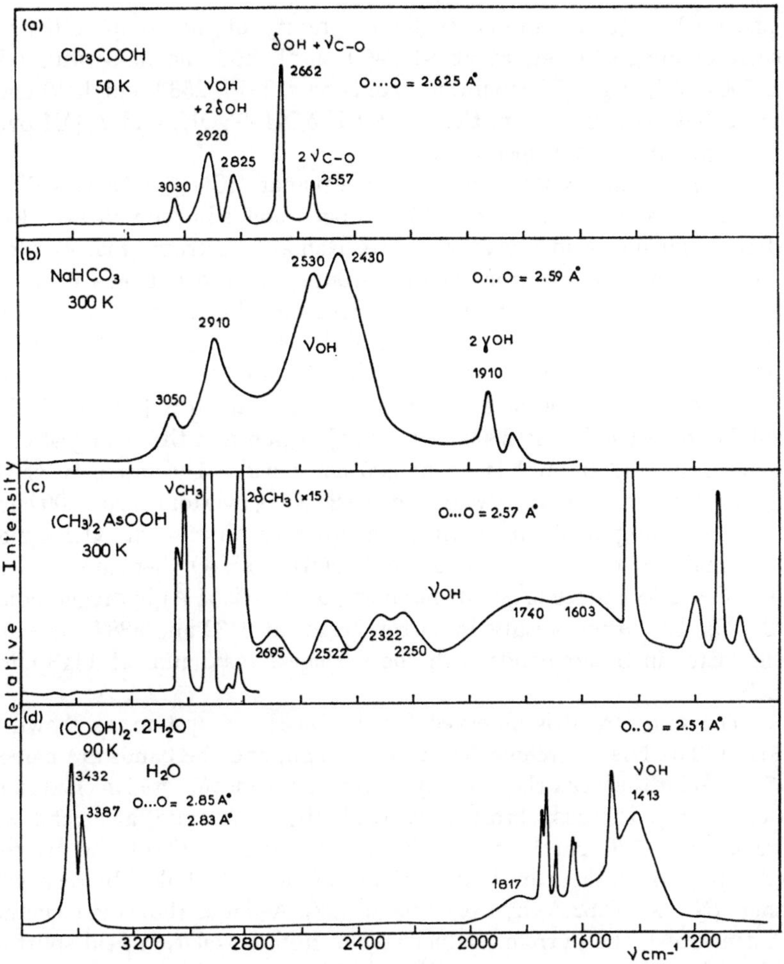

Fig. 4. Raman spectra of (a) CD_3COOH crystal at 50 °K; (b) sodium hydrogen carbonate at 300 °K; (c) dimethylarsinic acid $(CH_3)_2AsOOH$ at 300 °K; (d) oxalic acid dihydrate at 90 °K

νC—O, and 2νC—O overtone-combinations. The δOH and νC—O fundamentals are strongly coupled (*6*) and the above combination-overtones have thus all some OH stretching character (*17*). The OH stretching frequency corrected for resonances is estimated to be about 2870 cm^{-1}, and the half-width of the νOH band does not exceed 60 cm^{-1}. The sub-bands are thus much better resolved in the Raman than in the infrared spectrum.

The $NaHCO_3$ crystal contains infinite chains of HCO_3^- ions linked by shorter hydrogen bonds (O..O = 2.59 Å) (*18*) than those in acetic acid.

The νOH frequency of the center of gravity of the complex band is correspondingly lower, at about 2460 cm^{-1}, and the half-width ($\nu^1/_2 \simeq 300\ cm^{-1}$) larger. The main sub-bands near 2910, 2530 and 1910 cm^{-1} are believed to arise from the to 2δOH, δOH + γOH, and 2γOH overtone-combinations, respectively.

There is one centrosymmetric non-planar $[(CH_3)_2AsOOH]_2$ dimer containing short (O..O = 2.57 Å) hydrogen bonds in the unit cell of the dimethylarsinic acid crystal (*19*). Its Raman spectrum (Fig. 4) shows narrow bands arising from methyl groups and broad bands from OH groups. The structure of the O—H stretching bands is again ascribed to overtone-combinations-mainly 2δOH, γOH + δOH, and 2γOH in Fermi resonance with the OH stretching vibration (*20*). The "unperturbed" frequency of the latter is likely to be near $2000 \pm 50\ cm^{-1}$. The sub-bands occur in the 2800—1300 cm^{-1} region and the strongest component near 1700 cm^{-1} is about 600 cm^{-1} wide. The intensity of the maxima, however, is very weak and the $2\delta_s CH_3$ overtone near 2800 cm^{-1} had to be amplified fifteen times in order to observe the OH bands.

Finally, in oxalic acid dihydrate, $(COOH)_2 \cdot 2H_2O$, there are two long (O..O = 2.85 and 2.83 Å) and one short (O..O = 2.51 Å) hydrogen bonds (*21,22*). The former results in narrow bands at 3432 and 3387 cm^{-1} and the latter in broad bands with the strongest maximum at 1413 cm^{-1} (*23*).

The same trend is observed for the N—H..X system, as shown in Fig. 5. The NH stretching frequency is high, and the bands are narrow (5 cm^{-1}) for the weak N—H..π hydrogen bonds encountered in crystalline pyrrole (*16*). There is a broad νNH band ($\nu^1/_2 \simeq 270\ cm^{-1}$) at much lower frequency, 2720 cm^{-1}, in the Raman spectrum (*26*) of the triazole crystal in which there are infinite chains of molecules linked by relatively short (N..N = 2.82 Å) hydrogen bonds (*27*). As far as the band structure is concerned, the pyrrole doublet results from correlation field splitting while the submaxima of the NH stretching band of triazole arise from combination-overtones.

Hydrazinium chloride, which contains NH^+..Cl^-(N..Cl = 3.12 Å) and N—H^+..N (N..N = 2.95 Å) hydrogen bonds (*25*), is intermediate in frequencies (νNH = 2982, 2963 and 2920 cm^{-1}) [1] in half width ($\nu^1/_2 \simeq 20\ cm^{-1}$), and in structure. The sub-bands arise from intra- and intermolecular coupling as well as from combinations (*24*).

The behavior of an AH stretching band of a hydrogen bonded system in the Raman spectrum is thus in many ways similar to that in the infrared as far as frequency, band-width, and structure are concerned,

---

[1]) NH stretching frequencies observed in the spectrum of an isotopically mixed crystal containing 3% of $HD_4N_2^+$ (*24*).

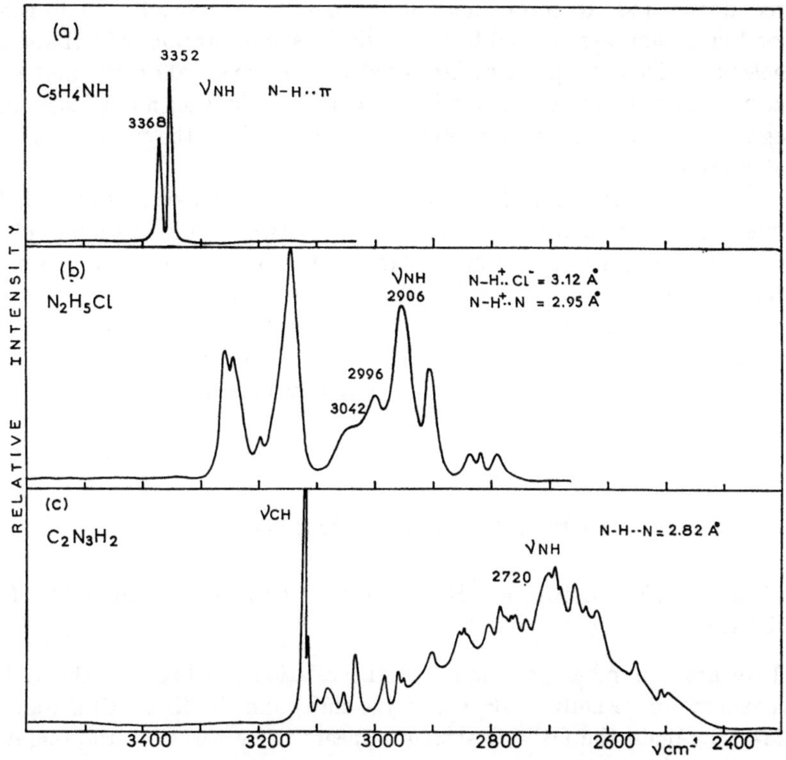

Fig. 5. Low temperature (90 °K) Raman spectra of polycrystalline (a) pyrrole, C_5H_4NH; (b) hydrazinium chloride N_2H_5Cl; (c) triazole, $C_2N_3H_2$

although the νAH Raman bands are usually less broad. The intensity of an AH stretching fundamental, on the other hand, is quite weak in the Raman but very strong in the infrared spectrum. The weakness of the νAH Raman band probably explains the scarcity of Raman data on hydrogen bonded systems. Yet Raman studies at low temperature of single crystals containing hydrogen bonds appear promising.

We have thus far discussed the A—H stretching fundamental. However, in a real system the A—H..B complex contains generally many more than three atoms and the hydrogen bond may modify the frequency, intensity, and shape of a number of bands. Nevertheless, the AH group vibrations (stretching and bending modes) usually remain the most sensitive, particularly when they are not mixed much with other vibrations. In addition to these changes, the formation of an A—H..B complex gives rise to new vibrations which do not exist for free donor and acceptor molecules. These intermolecular hydrogen-bond vibrations are derived from translations and rotations of AH and B molecules. Their frequencies are usually lower than 300 cm^{-1} and the most interesting one appears to be the ν(AH)..B hydrogen bond stretching frequency which can give direct information about the strength of the interaction studied.

In order to examine the A—H..B hydrogen bond strength we shall thus study the correlations that exist between the intramolecular A—H frequencies and intermolecular ν(AH)..B frequencies on the one hand and the r(A—H) and R(A..B) distances on the other.

The hydrogen bond type (*i.e.*, the qualitative potential curve) can be investigated by studying the νAH/νAD isotopic frequency ratio as a function of the intermolecular distance and νAH frequency.

II. Correlation of Frequencies and Distances

1. Relationship between AH Stretching Frequency and R(A..B) Distance

There are a number of studies on the relationship between the νAH frequency (or $\Delta\nu$ shift or $\Delta\nu/\nu_0$ relative shift) and the R(A..B) distance (*28—37*). Again the OH..O system appears to be the most thoroughly investigated.

We have re-examined this problem (Tables 2 and 3, Figs. 6 and 7), since in most of the earlier correlations there was some uncertainty concerning the A—H stretching frequencies and R(A..B) distances. This undoubtedly complicated the relationship which should be a simple one, at least for linear hydrogen bonds. *Ratajczak* and *Orville—Thomas* (*37*) remarked that for weak O—H..O hydrogen bonds there is an excellent correlation while for stronger hydrogen bonds the problem is complicated by the fact that more than one broad band (some with fine structure) is observed and the origin of these bands is not fully understood. Furthermore, the accuracy in determination of certain distances has been improved considerably. The R(O..O) distance in the crystal of potassium hydrogen bisphenylacetate, $KH(C_6H_5CH_2COO)_2$, for instance, was 2.55 ±0.1 Å in 1949 (*111*) and was revised by the same author to 2.440 ±0.004 Å in 1968 (*96*). The R(O..O) distance in the acetic acid crystal has been revised from 2.61 Å (*112*) to 2.63 Å, the latter value determined by both X-ray (*10*) and neutron (*9*) diffraction methods. The R(O..O) distance in the acid dimer in the gas phase, on the other hand, has been revised from 2.76 Å (*113*) to 2.68 Å (*67*).

Table 2. R(O...O) distances and OH stretching frequencies of some compounds containing O—H...O hydrogen bond

No.	Compound	R(O...O) (A°)	Ref.	νOH (cm^{-1})	Ref.
1	$MgSO_4 \cdot 6H_2O$	2.90	(*38*)	3520	(*39*)
		2.80		3455	
		2.78		3414	
		2.77		3395	
		2.76		3383	
		2.72		3320	
		2.70		3203	
2	$MgHPO_3 \cdot 6H_2O$	2.80	(*40*)	3420	(*39*)
		2.65		3127	
3	$MgK_2(SO_4)_2 \cdot 6H_2O$	2.75	(*41*)	3332	(*39*)
		2.74		3290	
		2.67		3250	
		2.66		3224	
		2.65		3155	

Table 2. (continued)

No.	Compound	R(O...O) (A°)	r(O—H) (A°)	Ref.	νOH (cm^{-1})	Ref.
4	$CaSO_4 \cdot 2H_2O$	2.82 2.81	(0.981) (1.002)	(*42*)	3494 3404	(*43*)
5	$NaHC_2O_4 \cdot H_2O$	2.83 2.81		(*44*)	3413 3410	(*14*)
6	$NaHSO_4 \cdot H_2O$	2.90 2.84		(*46*)	3480 3480	(*47*)
7	$(COOH)_2 \cdot 2H_2O$	2.86 2.88	(0.956) (0.964)	(*21, 22*)	3480 3420	(*48*)
8	Ice VI	2.81		(*49*)	3338	(*50*)
9	Ice I	2.76	(1.01)	(*51, 52*)	3277	(*53*)
10	H_3BO_3	2.72		(*54, 55*)	3200	(*56*)
11	C_6H_5OH	2.69		(*57*)	3225	(*58*)
12	CH_3OH	2.66		(*59*)	3245	(*60*)

Table 2 (continued)

No.	Compound	R(O...O) (A°)	r(O—H) (A°)	Ref.	νOH (cm^{-1})	Ref.	γOH (cm^{-1})	Ref.
13	$\alpha(COOH)_2$	2.71		(*61*)	3125	(*48*)	796	(*48*)
14	$KHSO_4$	2.68		(*45*)	2820	(*47*)		
	$NaHSO_4$	2.66		(*46*)	2890	(*47*)		
15	C_6H_5COOH	2.64		(*62*)	2900	(*63*)	920	(*63*)
16	$(CH_2COOH)_2$	2.64		(*64*)	2800	(*63*)	926	(*63*)
17	H_2SO_4	2.63		(*65*)	3030	(*66*)	620	(*66*)
18	CH_3COOH (crystal)	2.63	(1.011)	(*9*)	2875	(*6*)	923	(*6*)
	CH_3COOH (dimer)	2.684	(1.03)	(*67*)	3027	(*68*)	940	(*68*)
19	$KHCO_3$	2.61	(1.03)	(*69, 70*)	2630	(*71*)	980	(*71*)
20	$NaHCO_3$	2.59	(1.08)	(*18, 70*)	2540	(*71*)	998	(*71*)
20a	$RbHCO_3$	2.56	(1.10)	(*70*)	2570	(*145*)	1000	(*145*)
21	HCOOH (dimer)	2.703	(1.036)	(*72*)	3110	(*73*)	917	(*73*)
	HCOOH (crystal)	2.58		(*74*)	2650	(*73*)	974	(*73*)
22	$(CH_3)_2AsOOH$	2.57		(*19*)	2000	(*20*)	1003	(*20*)

Table 2 (continued)

No.	Compound	R(O...O) (A°)	r(O—H) (A°)	Ref.	νOH (cm⁻¹)	Ref.	γOH (cm⁻¹)	Ref.
23	$NaHC_2O_4 \cdot H_2O$	2.57		(*75*)	1800	(*14*)	1042	(*14*)
24	KHC_2O_4	2.523	(1.054)	(*76, 77*)	1500	(*15*)	1120	(*15*)
25	$NH_4H(CH_3COO)_2$	2.52 2.47		(*78*) (*78*)	1700 900	(*7*) (*7*)		
26	$(COOH)_2 \cdot 2H_2O$	2.51	(1.026) (1.057)	(*21*) (*22*)	1550	(*23*)	1117	(*23*)
27	$(CH_2COOH)_3N$	2.51		(*79*)	1400	(*80*)	1200	(*80*)
28	$KH(C_6H_5COO)_2$	2.51		(*81*)	1100	(*82*)	1220	(*82*)
29	$NaHCO_3Na_2CO_3 \cdot 2H_2O$	2.50 2.76 2.78	 (1.03) (0.99)	(*83, 84*)	1070 3030 3500	(*71*)	1215	(*71*)
30	$HCrO_2$	2.49	(1.24)	(*85*)	1650	(*86*)	(1229)	(*86*)
31	$LiHC_2O_4 \cdot H_2O$	2.49		(*87*)	1400	(*88*)	1100	(*88*)
32	$KH(CH_3COO)_2$	2.48		(*11*)	1400	(*7*)	1150	(*7*)

33	EDTA	2.46	(1.19)	*(89)*	800	*(90)*		
34	$N_2H_5HC_2O_4$	2.45	(1.225)	*(91)*	850	*(92)*	1218	*(92)*
35	$KHCH_2(CH_2COO)_2$	2.445	(1.22)	*(93)*	1000	*(82)*	1245	*(82)*
36	$NaH(CH_3COO)_2$	2.44	(1.22)	*(12)*	720	*(8)*	1285	*(8)*
37	$KH(CH_2COO)_2$	2.44	(1.22)	*(94)*	850	*(95)*	1350	*(95)*
38	$KH(C_6H_5CH_2COO)_2$	2.443	(1.22)	*(96)*	800	*(82)*	1270	*(82)*
39	$KH(CF_3COO)_2$	2.435	(1.217)	*(94)*	800	*(95)*		
40	CH_3COOH (monomer)		0.97	*(67)*	3583	*(5)*	650	*(97)*
41	CH_3COOH-Solution in following solvents							
	CH_3CN				3270	*(98)*	848[a]	
	$C_4H_8O_2$				3125	*(98)*	871[a]	
	C_4H_8O				3070[a]		891[a]	
	$CH_3CON(CH_3)_2$				2980[a]		905[a]	
	$(CH_3)_2SO$				2930[a]		925[a]	
42	$KHC_6H_4(COO_2)$				2000	*(82)*	1090	*(82)*
43	$NaHC_6H_4(COO)_2$				1500	*(82)*	1136	*(82)*
44	$KH(4\text{-}NO_2C_6H_4COO)_2$				2450	*(82)*	1015	*(82)*

[a] Values calculated from νOD and γOD frequencies.

As far as the determination of AH stretching frequencies is concerned we have established several criteria which are somewhat different for weak and for strong hydrogen bonds.

Weak Hydrogen Bonds: The νAH frequency data were obtained by using isotopically mixed crystals containing a small proportion, usually less than 5%, of the AH group in an AD environment. Under these conditions, the intermolecular (correlation field splitting) and intramolecular (for AH_2 groups) couplings are practically eliminated so that to each AH stretching frequency there corresponds an R(A..B) or *r*(A—H) distance.

There are multiple νNH bands in the infrared and Raman (Fig. 5) spectra of crystalline pyrrole. Only one band persists in the infrared spectrum of an isotopically mixed crystal and its frequency (3362 cm^{-1}) is very close to the average frequency (3360 cm^{-1}) of three infrared and two Raman bands. Incidentally, this shows that all pyrrole molecules in the crystal are symmetry equivalent and that there is only one N—H distance.

The infrared spectrum of an isotopically mixed crystal of $MgSO_4 \cdot 6H_2O$, on the other hand, shows seven νOH bands in the 3500—3200 cm^{-1} range (*39*) which correspond (Table 2) to seven R(O..O) distances between 2.90 and 2.70 Å (*38*).

The νAH stretching bands of isotopically mixed crystals at low temperature are usually quite narrow (5 to 15 cm^{-1}) and their frequencies can be determined with an accuracy of ± 2 cm^{-1} (± 0.03 to $\pm 0.06\%$).

Strong Hydrogen Bonds. The νAH bands of strong hydrogen bonds are much broader and frequently show some structure (Figs. 2, 3, 4 and 5). Assuming that all intensity arises mainly from the AH stretching fundamental, we can measure the frequency of the center of gravity of the band provided that the sub-bands are poorly resolved (as for the νNH band of triazole crystal (Fig. 5)). On the other hand, if the multiple bands are well resolved as in the Raman spectrum of CD_3COOH acid (Fig. 4), and if it can be shown that they result from Fermi resonances with the AH stretching fundamental, we can evaluate the unperturbed νAH frequency, supposing that the intensity of overtone-combination bands is negligible without resonant interactions (*114*). This assumption is not necessarily true for all systems, since the combinations having a certain A—H stretching character can be intense without resonances. This occurs, for example, for the 2δCH overtone of A_1 species of chloroform (*115*).

In centrosymmetric crystals, particularly for hydrogen-bonded dimers occuping sites of at least C_i symmetry, a knowledge of both infrared and Raman νAH frequencies is necessary: The monoclinic $KHCO_3$ crystal ($P2_1/a$) containing centrosymmetric cyclic $(HCO_3^-)_2$ dimers gives rise to an OH stretching vibration of B_u species, only infrared active, at 2680 cm^{-1} (*71*) and to one of A_g species, only Raman active at 2580 cm^{-1} (*116*). It is thus the average νOH frequency at 2630 cm^{-1} which must be ascribed to the 2.61 Å O..O distance. Similar considerations apply to the centrosymmetric crystals of $(COOH)_2 \cdot 2H_2O$ and $(CH_3)_2AsOOH$ (Table 2).

Generally, the uncertainty in determining the AH stretching frequency increases with the width and complexity of the band; for strong hydrogen bonds it can amount to ± 50 cm^{-1} ($\pm 3\%$) and even more. The low temperature study of crystals in the infrared and in the Raman can improve the accuracy of the band position considerably. However, crystallographic data about R(A..B) distances at low temperature are sparse, and the comparison of frequency vs. distance is thus less reliable.

a) O-H..O System. We have chosen, using the above criteria for frequencies, spectral and crystallographic data for about fifty-five O—H..O hydrogen bonds (Table 2) and have plotted the νOH frequency versus the R(O..O) distance in Fig. 6. The resulting curve shows that a very good correlation exists for weak hydrogen bonds and a reasonably good one for strong bonds. Two main characteristics of this curve should be mentioned.

1. The $\nu OH = fR(O..O)$ curve extends much further in frequency than did the previous ones, which were usually limited to 1600 cm^{-1}. A number of compounds studied recently which contain strong hydrogen bonds have νOH frequencies in the 1800—700 cm^{-1} range (*8,14,15,86, 92,95*). In particular, the centered hydrogen bonds of acid salts of carboxylic acids have OH stretching frequencies below 1000 cm^{-1} (*8,95*).

2. The steepness — $\Delta\nu/\Delta R$ — of the curve varies with the hydrogen bond strength. For the curve $\nu OH = f(RO..O)$ it increases from 1500 cm^{-1}/Å for long (2.90—2.70 Å) hydrogen bonds to 5000 and 12000 cm^{-1}/Å for intermediate (2.70—2.60 Å) and short (2.60—2.40 Å) hydrogen bonds, respectively. This is very significant, not only for the justification of the classification of hydrogen bonds, but above all for understanding the OH stretching band-width. We must point out that the band width (intrinsic band-width excluding the intermolecular coupling or structural disorder) increases an order of magnitude in going from weak ($\nu^1/_2 \simeq 10$ cm^{-1}) to medium-strong ($\nu^1/_2 \simeq 10^2$ cm^{-1}) hydrogen bonds and can

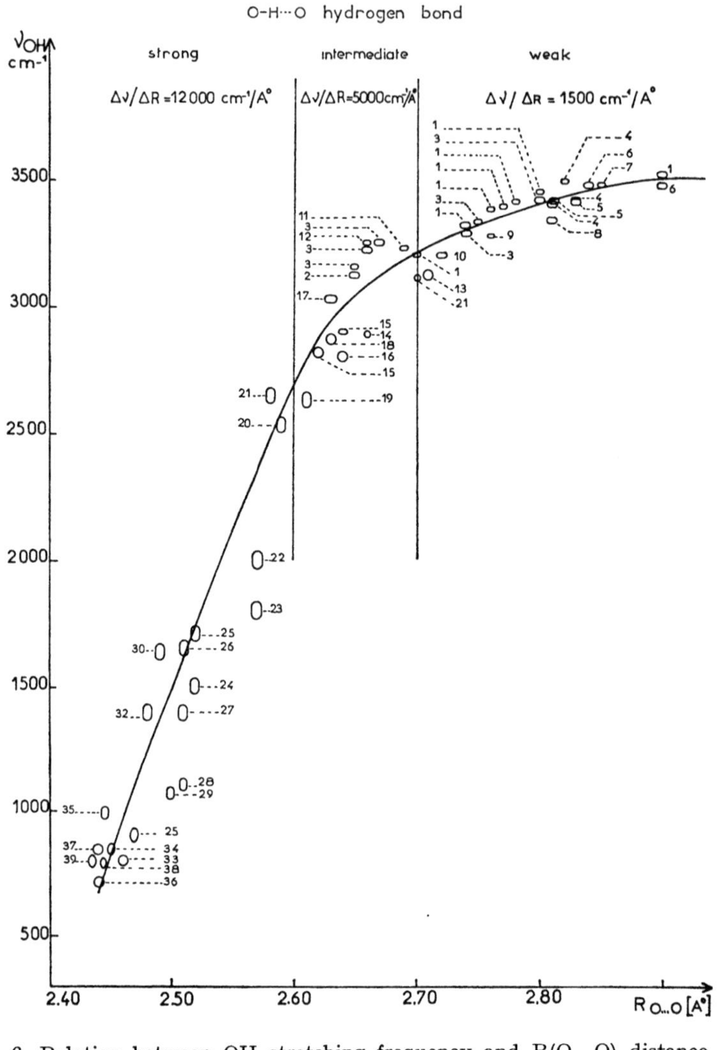

Fig. 6. Relation between OH stretching frequency and R(O..O) distance. The numbers are those of Table 2

amount to $\approx 10^3$ cm^{-1} for strong O—H..O bonds (Fig. 2). The $\Delta\nu/\Delta R$ factor, which increases in the same direction, appears to be the most important parameter determining the νAH band-width in hydrogen bonded systems (*117*).

b) N-H..N System. The NH stretching frequency versus R(N..N) distance plot for the N—H..N system is shown in Fig. 7. The correlation

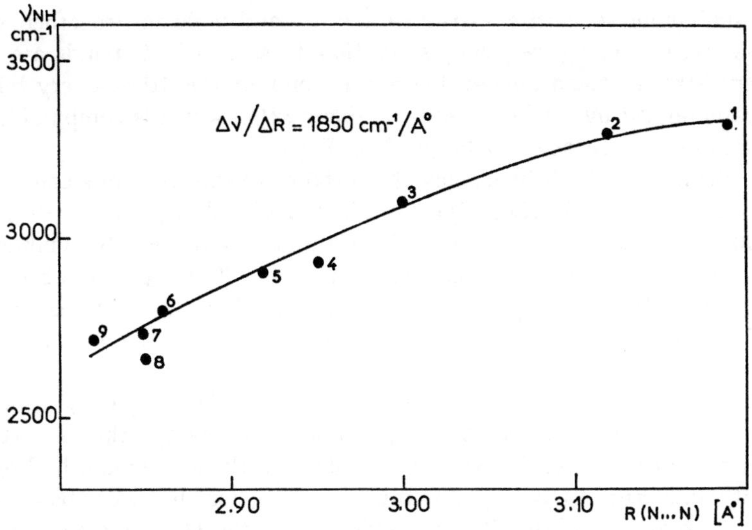

Fig. 7. Relation between NH stretching frequency and R(N..N) distance. The numbers are those of Table 3

Table 3. R(N...N) distances and NH stretching frequencies of some compounds containing the N—H..N hydrogen bond

No.	Compound	R(N...N) A°	(Ref.)	νNH frequency (cm^{-1})	(Ref.)
1	Hydrazine	3.19	(*99*)	3317	(*100*)
2	Formamidoxime	3.12	(*101*)	3300	(*102*)
3	9-methyladenine	3.00	(*103*)	3100	(*104*)
4	Hydrazinium chloride	2.95	(*105*)	2929	(*24*)
5	IMT—9 MA[a]) complex	2.92	(*106*)	2915	(*104*)
6	Imidazole	2.86	(*107*)	2800	(*108*)
7	Purine	2.85	(*109*)	2730	(*110*)
8	Hydrazinium hydrogen oxalate	2.85	(*91*)	2660	(*92*)
9	Triazole	2.82	(*27*)	2720	(*17*)

[a]) 1-methylthymine -9 methyladenine

appears satisfactory, but the available data are less numerous than for the O—H..O hydrogen bonds. The distinction between weak and strong hydrogen bonds is less pronounced: the steepness, $\Delta\nu/\Delta R = 1850\ cm^{-1}/Å$, is similar for the entire 3.20—2.80 Å range and indicates, together with the NH stretching frequency shift, that these N—H..N bonds are not very strong. The strongest hydrogen bond in the triazole crystal is characterized by a relative shift of about 22%, which is comparable to that in the acetic acid crystal (n° 18 in Fig. 6).

Strong N—H..N bonds have been shown to exist in compounds such as $HAu(CN)_2$ and $HCo(CN)_6$, for which νNH absorption bands near 1400 cm^{-1} are observed (*118*). Even stronger and possibly symmetric N—H..N hydrogen bonds are found in $(C_6H_5)_3AsH(NCS)_2$ and $(C_6H_5)_3PH(NCS)_2$ crystals with NH stretching frequencies below 1000 cm^{-1} (*119*). However, the crystal structures of these compounds are unknown.

If one believes in such correlation then the frequency-distance curves can be used for checking distances, at least within a certain range. For instance, the R(O..O) = 3.02 Å in the ammonium hydrogen phthalate crystal (*120*) appears much too long for the low νOH = 1500 cm^{-1} frequency (Table 2); we would expect the O..O distance to be somewhere near 2.53 Å. The R(O..O) = 2.49 Å of potassium hydrogen di-p-nitrobenzoate (*121*), on the other hand, seems too short for the νOH band observed near 2500 cm^{+1} (*82*); it should be close to 2.58 Å. Crystalline sulfuric acid contains hydrogen bonds characterized by an R(O..O) distance of 2.63 Å (*65*); its νOH frequency at 3030 cm^{-1} (*66*), however, would suggest a longer distance of about 2.68 Å.

This type of curve can also be used in estimating relative values of intermolecular distances, specially for weak hydrogen bonds. In the isotopically mixed crystal of $MgSiF_6 \cdot 6H_2O$ containing about 4% of the OD compound, six OD stretching bands are observed in a rather narrow region between 2611 and 2594 cm^{-1} (*39*). These frequencies appear to be in much better agreement with long (about 3 Å) O..O distances than with short O—H..F distances, the only ones considered in the isomorphous $FeSiF_6 \cdot 6H_2O$ crystal (*122*). Furthermore, we suggest that there are at least six O..O distances differing from one another by about 0.002 to 0.005 Å. (*39*).

2. Relationship between O—H Stretching Frequency and r(O—H) Distance

Theoretically the AH stretching frequency shift of a weak hydrogen bond is expected to be approximately proportional to the elongation (Δr) of the A—H bond; *Sokolov* (*2*) obtained the relation $\Delta\nu/\nu_0 = 4.922\,\Delta r$, for instance.

The data on r(O—H) distances are less numerous than those on O..O distances. Nevertheless, they allow us to construct the diagram given in

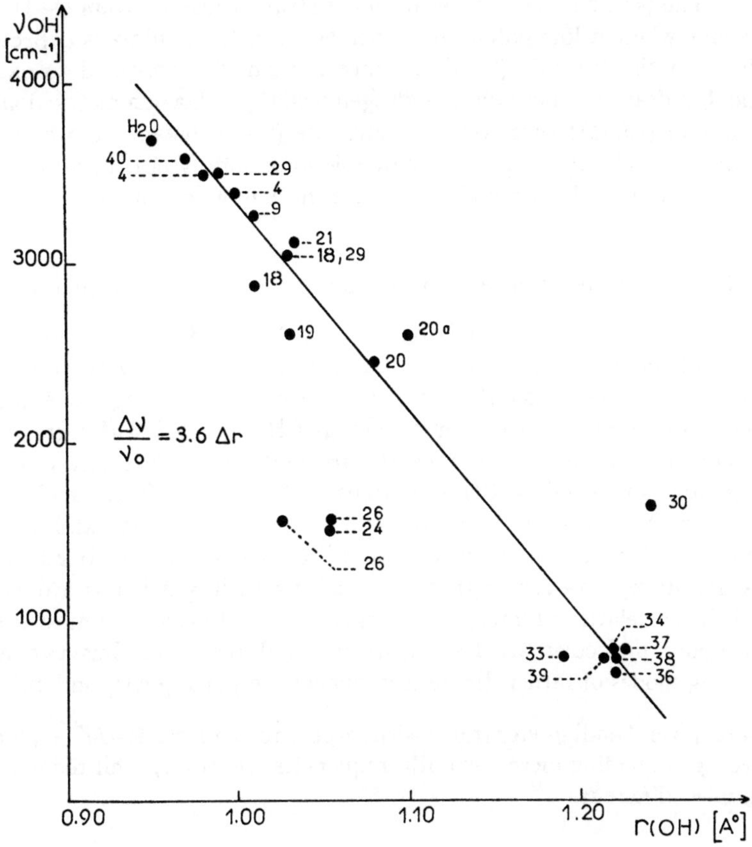

Fig. 8. OH stretching frequency as a function of *r*(O—H) distance for twenty-one *r*(O—H) distances. The numbers are those of Table 2 (*r*OH = 0.96 Å for the free water molecule)

Fig. 8. Three important points of this experimental relationship should be mentioned.

1. Most of the points corresponding to weak, intermediate, and strong hydrogen bonds are close to a straight line described by $\Delta\nu/\nu_0 = 3.6\ \Delta r$, which is not far from *Sokolov's* relation.

2. The $\Delta\nu$OH/ΔrOH parameter is about 11000 cm$_{-1}$/Å, *i.e.* the variation of the OH stretching frequency with the OH distance is about three times greater than for a bond such as C—O. For the latter system $\Delta\nu$CO/ΔrC—O is estimated from the values for CO, CO_2, CH_2O, HCOOH, CH_3COCH_3, and CH_3OH molecules (*123,124* and *125*) to be about 3300 cm^{-1}/Å.

3. The points lying farthest from the straight line are from O—H..O systems where a low potential barrier between the minima is expected. Those on the left side (O—H distance too small) correspond to oxalic acid dihydrate and potassium hydrogen oxalate with a symmetric double minimum (*15,48*); those on the right side (O—H distance too large to chromous acid with a symmetric double minimum (*4,86*). In all cases a delocalisation of the proton is likely (see the end of Section III).

3. Relationship between AH Stretching and Bending Frequencies

Model calculations for triatomic linear A—H..B systems show that the AH bending frequency should increase with decreasing AH stretching frequency (*2*). There is only one doubly degenerate bending mode in the above system and proton donors such as $CHCl_3$ or RC≡CH were used to look for correlations between the frequencies and intensities of the stretching and bending CH vibrations (*126,127,128*). These molecules are favorable since the CH oscillator remains degenerate and rather pure, *i.e.* there is not much mixing with other degrees of freedom; a disadvantage, however, is that the resulting hydrogen bonds are weak and the correlations limited to a narrow range. Moreover, in most cases the donor molecules are less symmetric and there are thus two AH bending modes of rather different frequency: in-plane (δAH) and out-of-plane (γAH) bending vibrations with respect to the local R—A/H plane. Hydrogen bonding increases both frequencies. However, their diagnostic value is different.

a) AH in-Plane Bending Vibration. OH and NH stretching modes are very rarely mixed with other internal degrees of freedom and remain localized within the respective group. This is much less so for in-plane bending OH and NH vibrations. They appear usually more or less strongly coupled with one or several other vibrations. Furthermore, the type and the strength of this vibrational interaction depends on the hydrogen bond strength.

The δOH mode in the CH_3COOH monomer is strongly mixed with the CH_3 symmetric bending and the C—O stretching modes (*5*) while in the crystal the coupling is limited mainly to the C—O stretch (*6*). Finally, in the sodium hydrogen diacetate crystal, which contains a much stronger hydrogen bond, the δOH deformation increases considerably in frequency; it is thus decoupled from the νC—O vibration and is mixed with the C=O stretch (*8*). In other words, we must compare the 1382—1264—1182 cm^{-1} triplet of the monomer with the 1418—1284 cm^{-1} and 1711—1542 cm^{-1} doublets of CH_3COOH and $NaH(CH_3COO)_2$ crystals, respectively. There are three broad bands at 1420, 1320, and 1275 cm^{-1}, which are sensitive to dilution effect, in the infrared spectrum of pure liquid ethanol. They have been assigned to coupled CH_2 scissoring, OH in-plane bending, and CH_2 twisting modes (*129*). The δOH vibration of p-chlorophenol appears to be heavily mixed with ν_{19b}, ν_{14} and ν_3 ring vibrations (*130*). The NH in-plane bending mode of molecules such as imidazole (*131*), triazole (*26*), purine (*110*), and pyrrole (*132*) is coupled with at least two or three heterocyclic ring vibrations. The effect of hydrogen bonding thus does not result in a large shift of one (δNH) band but shows up as a number of small frequency shifts (*26, 110*). Therefore, vibrational coupling makes the δAH frequency less specific, less sensitive, and, therefore, less suitable for correlation studies.

b) Out-of-Plane AH Bending Vibration. The γAH vibration frequently appears to be rather pure and sensitive towards molecular interactions. The corresponding absorption band (the γAH Raman bands are usually very weak) can thus be used as a criterion for hydrogen bond strength. It seems useful to distinguish two types of vibration. (1) Out-of-plane vibration of hydrogen bonded systems which transforms itself into a low-frequency torsional motions for the free monomer. This is the case of many alcohols and phenols, in particular, and is characterised by a big relative γAH frequency shift. The torsional frequency of gaseous methanol observed at 197 cm^{-1} (*133*) increases to 730 cm^{-1} when it becomes a γOH vibration of solid CH_3OH (60) the relative $(\gamma_0 - \gamma)/\gamma_0$ shift being about 270%. Much the same happens to the torsional vibration of phenol which increases from 306 cm^{-1} gas value (τ OH) to 720 cm^{-1} (γOH) for crystalline C_6H_5OH (58) and can reach a high frequency near 995 cm^{-1} in the *p*-nitrophenol-potassium acetate adduct (*134*). (2) γAH out-of-plane motion does not change its character when going from bonded to free species. This is the case of carboxylic acids and of heterocyclic molecules containing NH group.

For the first type a correlation between OH stretching and out-of-plane bending type of 1,2 disubstituted phenols has been shown to exist by *Nyquist* (*135*). For the second type we are proposing the correlations existing for some other O—H..O and N—H..X systems.

O—H..O system. When OH stretching frequency of twenty-eight compounds, mostly carboxylic acids and their acid salts, is plotted against corresponding γOH frequency the curve (a) in Fig. 9 is obtained. Majority of points are on a straight line with $\Delta\nu OH/\Delta\gamma OH = 5.8$.

In other terms, the variation of γOH frequency, between 800 and 1300 cm^{-1}, is almost six times less than that of OH stretching frequency, between 3800 and 700 cm^{-1}. The $\Delta\gamma/\gamma_0$ relative shift, on the other hand,

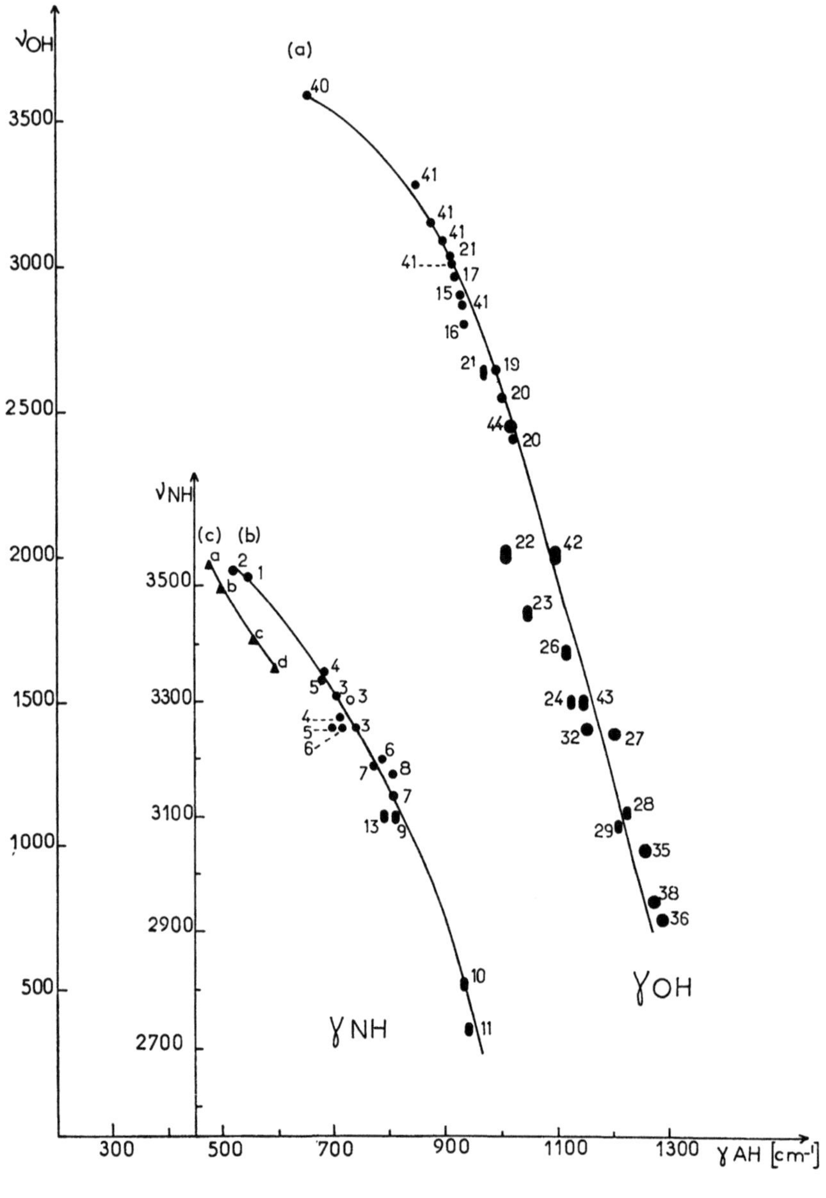

Fig. 9. Relation between AH stretching and out-of-plane bending frequencies; (a) νOH as a function of γOH frequency for twenty eight OH groups (numbers are those of Table 2); (b) νNH as a function of γNH frequency for seventeen NH groups (numbers are those of Table 4); (c) νNH and γNH frequencies of pyrrole (a) gas; (b) solution; (c) pure liquid and (d) solid at 90 °K

is always greater than the $\Delta\nu/\nu_0$ and can attain 100% for the shortest hydrogen bonds.

The analytical importance of γOH infrared bands appears to be considerable, especially for strong hydrogen bonds, where the corresponding νOH bands are very broad and the accuracy of the νOH frequency determination is limited to ± 50 cm^{-1} or more. The γOH bands, on the other hand, are quite narrow at liquid nitrogen temperature. The γOH frequency can under such conditions be assigned within ± 2 cm^{-1}. It should be noted that the γOH frequency is highly temperature-sensitive with a temperature shift (ν300 °K—ν90 °K)/ ν300 °K of several per cent.

N—H..X system. (X = Cl$^-$, Br$^-$, O, N). NH stretching and out-of-plane bending frequencies can also be simply related, as shown in Fig. 9b. The points (Table 4) correspond to frequencies of imidazole and triazole molecules in different environments, *i.e.*, physical state and in complexes with various metal salts. The range of the νNH (3550—2700 cm^{-1}) and γNH(515—940 cm^{-1}) frequencies is smaller than that of the O—H..O system, the N—H..X hydrogen bonds being weaker. The $\Delta\gamma$NH frequency shift is less than half that of the $\Delta\nu$NH shift and the $\Delta\gamma/\gamma_0$ relative shift for imidazole crystal amounts to 83%. As shown by deuteration studies of these molecules (*26, 131*) the NH out-of-plane mode of imidazole and triazole does not appear to be mixed with other out-of-plane vibrations. Yet the γNH(ν_{21}) vibration of pyrrole is strongly

Table 4. NH stretching and out-of-plane bending frequencies

N⁰	Compound[a]	νNH (cm^{-1})	γNH (cm^{-1})	N⁰	Compound	νNH (cm^{-1})	γNH (cm^{-1})
1	Triazole (gas)	3510	538	7	$Cu(Im)_4Cl_2$	3298	730
2	Imidazole (gas)	3518	515			3189	774
3	Zn $(Im)_2Cl_2$	3310	705			3130	803
		3250	740	8	$Cu(Im)_4Br_2$	3294	722
4	Zn $(Im)_2Br_2$	3350	685			3171	807
		3270	708	9	Ag $(Im)_2NO_3$	3100	810
5	Zn $(Im)_2I_2$	3337	682	10	Imidazole (crystal)	2800	937
		3250	695	11	Triazole (crystal)	2720	938
6	Cu $(Im)_2Cl_2$	3275	706	13	Triazole (liquid)	3100	790
		3199	783				

[a]) Data concerning the compounds n⁰ 2 and 11 correspond to Ref. (*108, 131*), n⁰ 1, 11 and 13 to Ref. (*26*) and n⁰ 3 and 9 to Ref. (*136*); Im = imidazole.

coupled to the ν_{23} skeletal mode of the same B_2 symmetry species (*132*). Nevertheless, the ν_{21} mode remains hydrogen-bond sensitive and its frequency increases from 473 to 500, 558, and 594 cm^{-1} in going from gas to carbon disulfide solution, pure liquid, and crystal at 90 °K, respectively (Fig. 9c). In the protonated nitrogen bases, such as pyridine (*137*), pyrimidine (*13*), and pyrazine (*138*) containing the NH^+ group (isoelectronic with the CH group), the γNH vibration mixes with two other out-of-plane modes which all become hydrogen-bond sensitive. The γNH, ν_{10b}, and ν_{11} frequencies of pyrimidinium ion, for instance, decrease about 6.5, 3 and 2.5%, respectively, when the chloride ion in $NH^+..X$ hydrogen bond is substituted by iodide (*13*).

4. Relationship between (AH)..B Stretching Frequency and R(A..B) Distance (or AH Stretching Frequency)

Strengthening of an A—H..B hydrogen bond implies lengthening of r(A—H) and shortening of R(A..B). The intramolecular F_{A-H} force constant is thus expected to decrease and the intermolecular $F_{H..B}$ force constant to increase under the same conditions. The determination of the F_{A-H} force constant from the AH stretching frequency is straightforward while the $F_{H..B}$ force constant is usually more difficult to estimate, particularly in crystals, where the ν(AH)..B vibration and other hydrogen-bond vibrations may couple with both internal and lattice modes. Hence we do not expect a simple relationship between ν(AH)..B and νAH frequencies or R(A..B) distance. A knowledge of the crystalline force field is necessary, and in most cases the diatomic model is not satisfactory (*139*).

However, there are some rigid systems with monoatomic acceptor groups such as $RNH^+..X^-$ hydrogen bonds of heterocyclic protonated bases (X = Cl, Br, I). In these compounds the ν(NH..X) vibration can be considered as an antitranslational motion which is not mixed with either internal or external modes (*140*). There is thus a simple relationship between ν(NH)..X and νNH stretching frequencies, and the diatomic model approximation for the $F_{H^+..X^-}$ force constant determination appears satisfactory. Furthermore, a plot of $F_{H^+..B^-}$ values as a function of $\Delta\nu/\nu_0$ (relative shift of the NH stretching frequency) provides an estimate of the relative contributions of ion-ion interaction and hydrogen bond to the force acting between the RNH^+ donor and X^- acceptor molecules (*140*).

There are too few diffraction data for the $NH^+..X^-$ system of protonated heterocycles. We thus had to look for correlations between the $F_{H..O}$ force constants and the R(O..O) distances (Fig. 10). The compounds chosen cover a rather wide range of R(O..O) distances, between

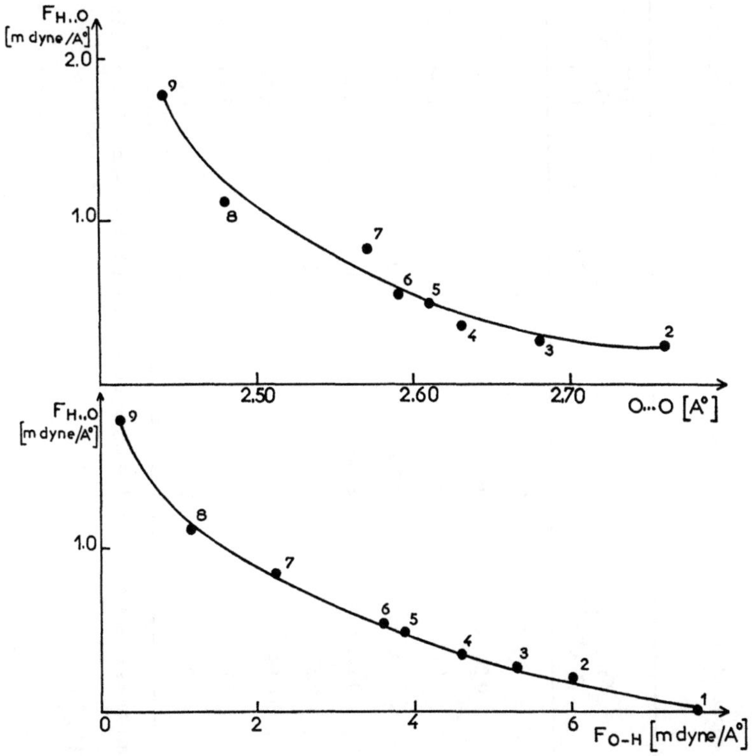

Fig. 10. Intermolecular hydrogen-bond force constant ($F_{H..O}$) as a function of R(O..O) distance and of intramolecular O—H force constant (F_{O-H}). Numbers are those of Table 5.

2.76 Å (ice) and 2.44 Å (sodium hydrogen diacetate). The $F_{H..O}$ force constants have been determined using the same diatomic model approximation (Table 5). This is certainly an oversimplification for these systems are less rigid than the protonated bases, and the OH..O stretching vibration does not necessarily correspond to a pure antitranslational motion (*8*). Thus the $F_{H..O}$ values are likely to be overestimated. Nevertheless, the plot of $F_{H..O}$ as a function of R(O..O) distance (Fig. 10a) shows the expected trend. There is even a better alignment of points in the plot of $F_{H..O}$ versus F_{O-H} (Fig. 10b); as usual correlation between spectroscopic data appears to be better than that between spectroscopic and diffraction data. It should be pointed out that the $F_{H..O}$ and F_{O-H} force constants are similar for a short (2.48 Å) OH..O bond in $KH(CH_3COO)_2$ and that $F_{H..O}$ is much greater than F_{O-H} for the symmetric hydrogen bond of sodium hydrogen diacetate.

Table 5. Correlation of the O—H...O hydrogen bond force constants ($F_{H...O}$)[a] with the O...O distances and the O—H bond force constants (F_{O-H})[b]

N°	Compound	ν(OH)...O Frequency[c] (cm^{-1})	Ref.	$F_{H...O}$ (mdyne/A°)	νOH Frequency (cm^{-1})	F_{O-H} (mdyne/Å°)	Distance[c] O...O (Å°)
1	H_2O gas				3707	7.60	
2	H_2O Ice Ih	220	*(53)*	0.25	3277	6.02	2.76
3	CH_3COOH gas	176	*(141)*	0.27	3027	5.30	2.68
4	CH_3COOH	198	*(142)*	0.35	2875	4.62	2.63
5	$KHCO_3$	234	*(116)*	0.49	2630	3.86	2.61
6	$NaHCO_3$	240	*(7)*	0.52	2540	3.60	2.59
7	$(CH_3)_2AsOOH$	203	*(20)*	0.84	2000	2.24	2.57
8	$KH(CH_3COO)_2$	250	*(7)*	1.09	1400	1.10	2.48
9	$NaH(CH_3COO)_2$	320	*(8)*	1.78	720	0.29	2.44

[a] $F_{H...O} = (4\pi^2 c^2 \nu^2 \mu)/nh$; $\nu = \nu$ (OH)...O, $\mu = m^2/2m$ (m: mass of the molecule 3, 4, 7) or of the anion 5, 6, 8, 9, n = 1 for 8 and 9; n = 2 for 3, 4, 5, 6, 7; see also Ref. *(141)*.
[b] $F_{O-H} = 4\pi^2 c^2 \nu^2 \mu$; $\nu = \nu OH$; μ = reduced mass of the O—H group.
[c] Values from Table 2.

III. Isotope Effect and the Position of the Proton. Correlation of νAH/νAD Isotopic Frequency Ratio and R(A..B) Distance

Strong O—H..O hydrogen bonds, usually characterized by an R(O..O) distance between 2.60 to 2.40 Å, may be divided approximately into asymmetric double minimum, symmetric double minimum and symmetric single minimum type.

All these types can exist in acid salts of carboxylic acids containing the (—COOHOOC—) group. Many of them have been studied by diffraction methods, particularly by *Speakman* and coworkers (*93* and references therein) and by *Olovsson* and his group (*10,44,75,87,91*). *Speakman* distinguishes A and B types of structures. The B type corresponds to asymmetric hydrogen bonds, *i.e.* the C—O distances and C—OO angles show clearly distinct COOH carboxylic and COO^- carboxylate groups of the complex anion $(RCOOHOOCR)^-$. The infrared and Raman spectra also show the bands characteristic of COOH and COO^- groups, respectively (*14,15,82,144*). Crystals of acid salts, *e.g.* $KH(4\text{-}NO_2C_6H_4COO)_2$, $KHC_6H_4(COO)_2$, $NaHC_2O_4 \cdot H_2O$, KHC_2O_4, $LiHC_2O_4 \cdot H_2O$ (Table 2) are good examples of type B. In the A type (*Speakman* (*93*) gives about fifteen examples among acid salts of mono- and dicarboxylic acids) the two carboxylate residues are related by a crystallographic symmetry element, a center of inversion, a twofold axis, or a mirror plane. There is thus a crystallographically symmetric hydrogen bond with equivalent RCOO groups. However, it does not appear possible using diffraction methods to distinguish a symmetric double minimum from a single minimum. Even neutron diffraction cannot discriminate between two minima separated by less than 0.1 Å and a single centered well (*Hamilton* and *Ibers* (*4*)). We shall therefore examine the spectroscopical method and its possibilities for distinguishing between these types of hydrogen bonding[2]).

Isotope Effect and Hydrogen Bond Type. Singh and *Wood* (*146*) have calculated using Stepanov's approximation the effect of isotopic substitution of hydrogen by deuterium on the R(A..B) distance of an A—H..B hydrogen bond. They found that the R(A..B) distance increases (positive isotope effect) on deuteration if the potential energy

[2]) Crystallographically symmetric hydrogen bonds may also be of a statistical nature. In the $RbHCO_3$ crystal for instance, HCO_3^- ions form cyclic dimers occupying C_{2h} site symmetry. The two-fold axis is in the dimer plane which implies symmetric hydrogen bonds (*70*). However, the high OH stretching frequency at 2570 cm^{-1} (*145*) eliminates the possibility of genuine symmetry; this crystal represents clearly a case of statistical symmetry (*70*).

function contains a symmetric or asymmetric double minimum. Similar conclusions have been reached by *Savelev* and *Sokolov* (*147*). On the other hand, the R(A..B) distance decreases (negative isotope effect) on deuteration if the potential curve has a single symmetric minimum. According to *Rundle* (*148*) a zero isotope effect is to be expected this case.

a) Double Minimum Type of Hydrogen Bond. Table 6 contains some data on the positive isotope effect ($\Delta R > O$) which varies from 0.0015 Å for Ih ice to 0.060 Å for $HCrO_2$. This effect would thus indicate that the corresponding hydrogen bonds are of the double minimum type. However, as *Speakman* (*143*) remarked, "The results of such studies have so far been disappointing, and this is because the effects are usually rather small and therefore need very precise analysis of both ordinary and deuterated crystals".

We have shown, on the other hand, that the positive isotope effect which implies a weakening of the hydrogen bond on deuteration, manifests itself in vibrational spectra by a decrease of the $\nu OH/\nu OD$ isotopical frequency ratio and by an increase of $\gamma OH/\gamma OD$ ratio (*8,14,15,71*). Table 6 and Fig. 11 illustrate this behavior: shortening of the R(O..O) distance is accompanied by a decrease of the OH stretching frequency and of the $\nu OH/\nu OD$ ratio. The value of this ratio is about 1.355 for free molecules and diminishes with increasing strength of the hydrogen bond, reaching values close to unity for crystals with the greatest positive isotope effect. The γOH frequency increases in the same way and is accompanied by a change of $\gamma OH/\gamma OD$ ratio from 1.32 for α-oxalic acid (anhydrous) to 1.46 for oxalic acid dihydrate[3]). The $\gamma OH/\gamma OD$ increase is less spectacular than that for stretching frequencies. However, it is significant and consistent with a weakening of the hydrogen bond on deuteration. This was recognized by Snyder and Ibers for $HCrO_2$ (*86*).

From the data of Table 6 for the O—H..O system and of Table 7 for the N—H..N system we conclude that the lowering of the $\nu AH/\nu AD$ ratio indicates a positive isotope effect and hence a double minimum type of hydrogen bond. Lower AH stretching frequency and $\nu AH/\nu AD$ ratio imply a greater isotope effect, in agreement with the predictions of *Singh* and *Wood* (*146*): ΔR increases with the lowering of the barrier between two potential wells, and a low barrier corresponds to a stronger hydrogen bond.

The special case of a symmetric double minimum occurs in the $HCrO_2$ crystal (*4,86*). The effective symmetry of the O—H..O hydrogen bond in this compound was shown by diffraction and spectroscopic

[3]) It should be emphasized that the high $\gamma OH/\gamma OD$ ratio is not the result of kinematic coupling since the γOH and γOD modes are rather pure.

Table 6. Correlation of the $\nu OH/\nu OD$ and $\gamma OH/\gamma OD$ isotopic frequencies ratios with their νOH and γOD frequencies and R(O...O) distances

Nº	Compound	$R_{O...O}$ Distance[a] (A°)	ΔR[b] ($A° \cdot 10^{-3}$)	Ref.	νOH Frequency[a] (cm^{-1})	$\frac{\nu OH}{\nu OD}$	γOH Frequency (cm^{-1})	$\frac{\gamma OH}{\gamma OD}$	Ref.
1	H_2O				3707	1.355			*(53)*
2	CH_3COOH monomer				3583	1.355	650		*(5,97)*
3	H_2O Ice I	2.76	+1.5	*(1)*	3277	1.350			*(53)*
4	$\alpha(COOH)_2$	2.71			3125	1.332	796	1.32	*(7)*
5	CH_3COOH crystal	2.625	(+10)	*(149)*	2875	1.310	923	1.36	*(6)*
6	$KHCO_3$	2.61			2630	1.28	980	1.38	*(71,116)*
7	$NaHCO_3$	2.59			2540	1.26	998	1.40	*(71)*
8	$(CH_3)_2AsOOH$	2.57			2000	1.1	1003	1.43	*(20)*
9	$NaHC_2O_4 \cdot H_2O$	2.57	+22	*(75)*	1800	1.0	1042	1.38	*(14)*
10	KHC_2O_4	2.523			1500	1.0	1120	1.39	*(15)*
11	$(COOH)_2 \cdot 2H_2O$	2.512	+19	*(21)*	1550	0.95	1168	1.46	*(88)*
12	$LiHC_2O_4 \cdot H_2O$	2.49	+16	*(87)*	1400	1.0	1101	1.39	*(88)*
13	$HCrO_2$	2.49	+60	*(85)*	1640	0.96	1202	1.44	*(86)*
14	$NaH(CH_3COO)_2$	2.444			720	1.44	1285	1.34	*(8)*
15	$KH(CF_3COO)_2$	2.437	0	*(150)*	800	1.43			*(95)*
16	$KH(CH_2COO)_2$	2.446			850	1.39	1300	1.33	*(95)*

a) Values from Table 2.
b) $\Delta R = R(O{-}D...O) - R(OH...O)$
c) Observed for the bi-molecule ($CF_3COOH{-}HCOOH$) *(149)* containing an O—H...O hydrogen bond similar to that of acetic acid.

data, but the deuterated derivative, $DCrO_2$, is no longer symmetric. This is interpreted in terms of a symmetric double minimum separated by a low barrier in $HCrO_2$. The barrier becomes high in $DCrO_2$ causing the proton to shift closer to one oxygen atom with a resulting asymmetric hydrogen bond (*4*). The isotope effect is the largest so far measured (0.06 Å) and the νOH band, a singlet near 1640 cm^{-1}, shifts to about 1750 cm^{-1} on deuteration and becomes a doublet. We agree with *Sheppard* (*151*) that the two components of the OD stretching band could be explained in terms of a Fermi resonance between νOD and 2δOD vibrations. From our criteria for frequency determination the νOH/νOD ratio is about 0.95 and is thus consistent with a large isotope effect and a double minimum.

A similar case, *i.e.* a symmetric hydrogen bond becoming asymmetric on deuteration, was predicted on the basis of infrared and Raman spectra for a crystallographically symmetric hydrogen bond in hydrazinium hydrogen oxalate, $N_2H_5HC_2O_4$ (*92*) and confirmed recently by diffraction methods (*156*).

b) The Single Symmetric Minimum Type of Hydrogen Bond. The best known centered hydrogen bond is that of the $(FHF)^-$ ion in the KHF_2 crystal (*4*). The antisymmetric and symmetric stretching frequencies have been observed at 1450 and 600 cm^{-1}, respectively, and the corresponding force constants are $F_a = 0.325$ mdynes/Å and $F_s = 4.08$ mdynes/Å. A shortening (negative isotope effect) of the R(F..F) distance by 0.0024 Å on deuteration has been found, in agreement with the calculated effect (*146*). The isotopic frequency ratio for the antisymmetric stretching mode is 1.414 (*152*).

The OHO symmetric hydrogen bonds are less clear cut although the existence of such bonds in the type A acid salts was noted many years ago. The crystallographic (*111,93*) and spectroscopic (*82,144,34*) data were in good agreement about the equivalence of the RCOO residues in the $(RCOOHOOCR)^-$ ion. However, the OH and (OH)..O stretching frequencies were uncertain or unknown and hence identification of the type of hydrogen bonding by spectroscopic means was not possible.

Recently, new spectroscopic data observed by studying low temperature infrared and laser Raman spectra of a few isotopic species of acid salts have enabled the identification of the bands of three type A acid salts, $KH(CF_3COO)_2$, $NaH(CH_3COO)_2$ and $KH(CH_2COO)_2$ (*8,95*).

It was shown that the selection rules derived from the C_i and C_2 site groups of the complex $(CF_3COOHOOCCF_3)^-$ and $(CH_3COOHOOCCH_3)^-$ ions, respectively, and from the C_{2h} factor group of the $(\text{—}CH_2COOHOOCCH_2\text{—})^-$ infinite chain of potassium hydrogen succinate were obeyed and thus the results were consistent with a symmetric hydrogen bond. This evidence is, however, insufficient to eliminate the possibility of a symmetric double minimum.

Sodium hydrogen diacetate and its deuterated derivatives, NaD $(CH_3COO)_2$, $NaH(CD_3COO)_2$, and $NaD(CD_3COO)_2$, present a particularly favorable case as far as the assignment of OH and OD bands is concerned. The νOH, δOH, and γOH frequencies have been assigued to the bands at 720, 1540, and 1285 cm^{-1}, respectively, and the νOD, δOD, and γOD analogues, at 510, 1150 and 960 cm^{-1}. This is one of the rare instances where the bending frequencies are higher than the stretching ones (Fig. 9). Furthermore, the symmetric (ν_sOHO) hydrogen bond stretching frequency was identified in the Raman at 320 cm^{-1}.

On the basis of these data we can conclude:

1. the OH stretching frequency at 720 cm^{-1} is very low and is consistent with a symmetric hydrogen bond since r(O—H) $= {}^1/_2$ R(O..O) $= 1.22$ Å is a very long distance (Fig. 8).

2. the isotopic frequency ratio, νOH/νOD $= 1.41$, is much higher than that of strong hydrogen bonds of the double minimum type (Table 6) and even higher than the νOH/νOD ratio of a free molecule. The γOH/γOD ratio of 1.34, on the other hand, is lower than that of double minimum hydrogen bonds in spite of a higher OH out-of-plane bending frequency. Isotopic substitution of acid hydrogen by deuterium thus makes the hydrogen bond stronger or at least does not weaken it, in agreement with theoretical predictions for a single central minimum (*146*).

3. The intermolecular $F_{H..O}$ force constant is almost an order of magnitude larger than the intramolecular F_{O-H} force constant (Table 5, Fig. 10), in much the same way as for KHF_2 crystal.

Similar conclusions (*95*) have been reached for $KH(CF_3COO)_2$ and $KH(CH_2COO)_2$ (Table 6) for which some additional data supporting hydrogen bond symmetry should be mentioned. The R(O..O) distance in the $KH(CF_3COO)_2$ crystal does not expand on deuterium substitution (Table 6) according to diffraction studies (*150*); the deuteron quadrupole coupling constant in crystalline $KD(CF_3COO)_2$ is very low (55.5 KHz) and similar to that of potassium deuterium maleate where a truly symmetric hydrogen bond (*153*) is believed to exist. The OHO hydrogen bonds in crystals of sodium hydrogen diacetate, potassium hydrogen di-trifluoroacetate, and potassium hydrogen succinate are thus genuinely symmetric and of the single central minimum type.

In Fig. 11 the νOH/νOD frequency ratio is plotted as a function of R(O..O) distance or OH stretching frequency. The resulting curve shows that the $\varrho = \nu$OH/νOD value ($\varrho = 1.355$ for free molecules) decreases slowly with shorter distances or lower frequencies for weak and medium strong hydrogen bonds (*i.e.* those longer than about 2.60 Å). The poten-

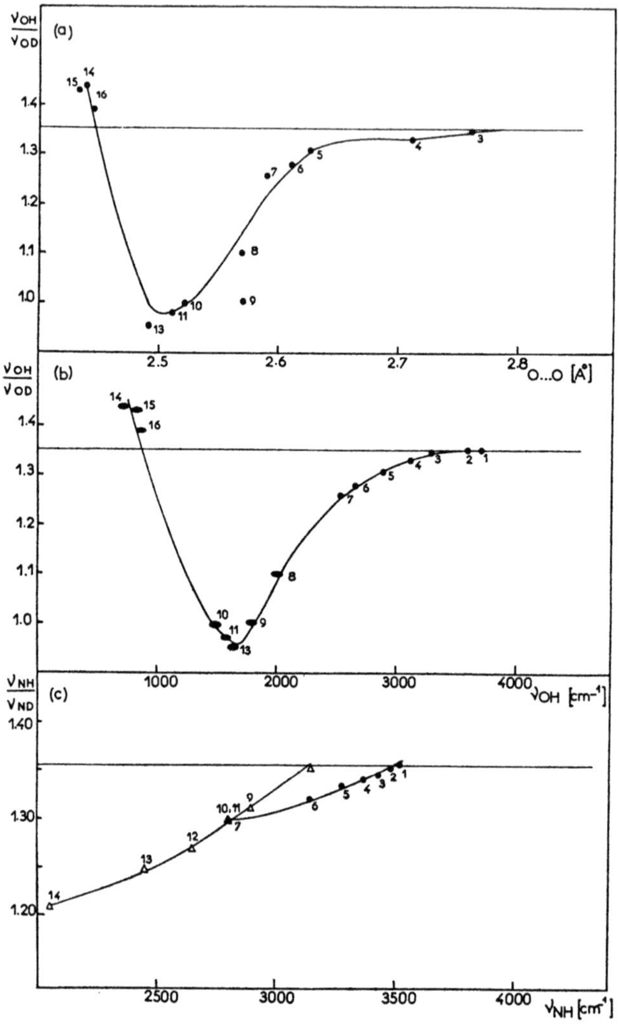

Fig. 11. Relation between νAH/νAD isotopic frequency ratio and νAH frequency or R(O..O) distance; (a) νOH/νOD ratio as a function of R(O..O) distance; (b) νOH/νOD ratio as a function of νOH frequency; (c) νNH/νND ratio as a function of νNH frequency. The numbers for the O—H..O and N—H..N systems are those of Table 6 and 7, respectively

tial barrier separating the two wells probably remains relatively high and the proton stays preferentially in one well. When the R(O..O) distance becomes shorter than about 2.60 Å the νOH/νOD ratio decreases rapidly; values of about 1 ± 0.1 are observed for R(O..O) distances between 2.58 and 2.48 Å. This range may correspond to a relatively low barrier and thus to a rather delocalized proton. As already pointed out in Section II the greatest deviations from the straight line $\Delta\nu/\nu_0 = 3.6\ \Delta r$ of Figure 8 are observed for the points in the 2.58—2.48 Å distance range (2000—1400 cm^{-1} frequency region). More diffraction and spectroscopic data on such systems would be very useful. Finally, the νOH/νOD ratio increases suddenly to values higher than 1.355 of free molecules for distances less than 2.46 Å. This range corresponds to a single central minimum type of hydrogen bond.

The νNH/νND isotopic frequency ratio of imidazole hydrogen bonded NH and NH^+ groups of some protonated nitrogen bases containing $NH^+..X^-$ hydrogen bonds (Table 7) also decreases with decreasing

Table 7. NH stretching frequencies and νNH/νND isotopic frequency ratios of the imidazolic NH group depending on the environment (1 to 7) and of the NH^+ group of some protonated nitrogen bases depending on the anion (8 to 14)

N°	Compound	νNH (cm^{-1})	νNH/νND	Ref.
1	Imidazole gas	3518	1.355	(*108*)
2	Imidazole in CCl_4	3480	1.352	(*154*)
3	Imidazole in C_6H_6	3434	1.345	(*154*)
4	Imidazole in CH_3CN	3365	1.341	(*154*)
5	Zn $(Im)_2Cl_2$ crystal	3280	1.335	(*136*)
6	Imidazole in DMSO	3145	1.320	(*154*)
7	Imidazole crystal	2800	1.305	(*108*)
8	Pyrimidinium Hexachlorostannate	3150	1.35	(*13*)
9	Pyrazinium bromide	2900	1.31	(*138*)
10	Pyrazinium chloride	2800	1.30	(*138*)
11	Pyridinium bromide	2800	1.30	(*137*)
12	N Methyl imidazolium chloride	2650	1.27	(*154*)
13	Pyridinium chloride	2450	1.25	(*137*)
14	N-Methyl diphenyl ammonium chloride	2050	1.21	(*155*)

NH stretching frequency. Both systems yield a simple relation between the isotopic frequency ratio and NH stretching frequency (Fig. 11c). However, unlike the O—H..O system, the νNH/νND ratio is at least 1.30 for imidazole and 1.21 for the protonated bases. This indicates that NH..Y and $N—H^+..X^-$ hydrogen bonds are not very strong and that the proton remains attached to the nitrogen even for the strongest interaction.

IV. Concluding Remarks

Complementarity of spectroscopic and diffraction methods appears particularly valuable for the study of hydrogen bonded crystals. Correlation of AH stretching and bending frequencies with R(A..B) distances not only gives a better idea about hydrogen bond strength but also shows the importance of the $\Delta\nu/\Delta R$ parameter as a factor determining AH stretching band breadth. Variation of the νOH frequency with O—H distance is about three times greater than that of the C—O bond. Correlation of νAH/νAD frequency ratio with R(A..B) distance of νAH frequency can give information about the isotope effect and hydrogen bond type. In particular the isotopic frequency ratio value may be decisive in distinguishing between the symmetric single minimum and the double symmetric minimum. Further correlations, such as AH stretching band intensity and shape (Raman and infrared) with rAH and R(A..B) distances, which may lead to a knowledge of the charge density distribution in an A—H..B hydrogen bond as a function of its strength, appear potentially interesting but experimental data are lacking.

Acknowledgment. I am grateful to all research associates who participated in hydrogen bonding studies.

V. References

1. *Pimentel, G. C., McClellan, A. L.:* Hydrogen Bond. San Francisco and London: W. H. Freeman and Co, 1960.
   — — Ann. Rev. Phys. Chem. *22*, 347 (1971).
2. *Sokolov, N. D.:* Vodorodnaya Svyaz', p. 7. Moscow 1964.
3. *Hadži, D.:* Boll. Sci. Fac. Chim. Ind. Bologna *21*, 23 (1963).
4. *Hamilton, W. C., Ibers, J. A.:* Hydrogen Bonding in Solids. New York: W. A. Benjamin Inc. 1968.
5. *Haurie, M., Novak, A.:* J. Chim. Phys. *62*, 137 (1965).
6. — — Spectrochim. Acta *21*, 1217 (1965).
7. Unpublished results of this laboratory.
8. *Novak, A.:* J. Chim. Phys. *69*, 1615 (1972).
9. *Jonsson, P. G.:* Acta Cryst. *27*, 893 (1971).
10. *Nahringbauer, I.:* Acta Chem. Scand. *24*, 453 (1970).
11. *Currie, M.:* J. Chem. Soc. Perkin II *1972*, 832.
12. *Speakman, J. C., Mills, H. H.:* J. Chem. Soc. *1961*, 1164.
13. *Foglizzo, R., Novak, A.:* Spectrochim. Acta *26A*, 2281 (1970).
14. *De Villepin, J., Novak, A.:* Spectrochim. Acta *27A*, 1259 (1971).
15. — — Spectry. Letters *4*, 1 (1971).

16. *Lautié, A., Novak, A.:* J. Chem. Phys. *56*, 2479 (1972).
17. *Foglizzo, R., Novak, A.:* Advances in Raman Spectroscopy I (ed. *J. P. Mathieu*), p. 515. London: Heyden and Son Ltd. 1972.
18. *Sass, R. L., Scheuerman, R. F.:* Acta Cryst. *15*, 77 (1962).
19. *Trotter, J., Zobel, T.:* J. Chem. Soc. *1965*, 4466.
20. *Hadži, D., Novak, A.:* to be published.
21. *Delaplane, R. G., Ibers, J. A.:* Acta Cryst. *B25*, 2423 (1969).
22. *Sabine, T. M., Cox, G. W., Craven, B. M.:* Acta Cryst. *B25*, 2437 (1969).
23. *De Villepin, J., Novak, A.:* to be published.
24. — — Mol. Cryst. (Liq. Cryst.) in press.
25. *Sakurai, K., Tomiie, Y.:* Acta Cryst. *5*, 289 (1952).
26. *Saint-Roch, B.:* Thesis, University of Bordeaux, 1972.
27. *Goldstein, P., Ladell, J., Abowitz, G.:* Acta Cryst. *25*, 135 (1969).
28. *Rundle, R. E., Parasol, M.:* J. Chem. Phys. *20*, 1487 (1952).
29. *Lord, R. C., Merrifield, R. E.:* J. Chem. Phys. *21*, 166 (1953).
30. *Nakamoto, K., Margoshes, M., Rundle, R. E.:* J. Am. Chem. Soc. *77*, 6480 (1955).
31. *Pimentel, G. C., Sederholm, C. H.:* J. Chem. Phys. *24*, 639 (1956).
32. *Glemser, O., Hartert, E.:* Z. Anorg. Allgem. Chem. *283*, 111 (1956).
33. *Lippincott, E. R., Schroeder, R.:* J. Chem. Phys. *23*, 1099 (1955).
34. *Blinc, R., Hadži, D., Novak, A.:* Elektrochem. *64*, 567 (1960).
35. *Schwarzmann, E.:* Z. Anorg. Allgem. Chem. *317*, 176 (1962).
36. *Wall, T. T., Hornig, D. F.:* J. Chem. Phys. *43*, 2079 (1965).
37. *Ratajczak, H., Orville-Thomas, W. J.:* J. Mol. Struct. *1*, 449 (1967—68).
38. *Zalkin, A., Ruben, H., Templeton, D. H.:* Acta Cryst. *17*, 235 (1964).
39. *Guillermet, J., Novak, A.:* J. Mol. Struct. *17*, 91 (1973).
40. *Corbridge, D. E. C.:* Acta Cryst. *9*, 991 (1956).
41. *Kannan, K. K., Viswamitra, M. A.:* Z. Krist. *122*, 161 (1965).
42. *Atoji, M., Rundle, R. E.:* J. Chem. Phys. *29*, 1306 (1958).
43. *Seidl, V., Knop, O., Falk, M.:* Can. J. Chem. *47*, 1361 (1969).
44. *Tellgren, R., Olovsson, I.:* J. Chem. Phys. *54*, 127 (1971).
45. *Cruickshank, D. W.:* Acta Cryst. *17*, 682 (1964).
46. *Grimvall, S.:* Acta Chem. Scand. *25*, 3213 (1971).
47. *Goypiron, A., De Villepin, J.:* to be published.
48. *De Villepin, J., Novak, A.:* to be published.
49. *Kamb, B.:* Science *150*, 205 (1965).
50. *Bertie, J. E., Labbe, H. J., Whalley, E.:* J. Chem. Phys. *49*, 2141 (1968).
51. *Barnes, W. H.:* Proc. Roy. Soc. (London) *125A*, 670 (1929).
52. *Peterson, S. W., Levy, H. A.:* Acta Cryst. *10*, 70 (1957).
53. *Whalley, E.:* Developments in Applied Spectroscopy, Vol. 6 p. 277. Plenum Press 1968.
54. *Cowley, J. M.:* Acta Cryst. *6*, 516 (1953).
55. *Zachariasen, W. H.:* Acta Cryst. *7*, 305 (1954).
56. *Bethell, D. E., Sheppard, N.:* Trans. Faraday Soc. *51*, 9 (1955).
57. *Scheringer, C., Wehrahn, O. J., Von Steckelberg, M.:* Z. Elektrochem. *64*, 381 (1960).
58. *Evans, J. C.:* Spectrochim. Acta *16*, 1382 (1960).
59. *Tauer, K. J., Lipscomb, W. N.:* Acta Cryst. *5*, 606 (1952).
60. *Falk, M., Whalley, E.:* J. Chem. Phys. *34*, 1554 (1952).
61. *Cox, E. G., Dougill, M. W., Jeffrey, G. A.:* J. Chem. Soc. *1952*, 4859.
62. *Sim, G. A., Robertson, J. M., Goodwin, T. H.:* Acta Cryst. *8*, 157 (1955).
63. *Hadži, D., Sheppard, N.:* Proc. Roy. Soc. London *216A*, 247 (1953).

64. *Morrison, J. D., Robertson, J. M.:* J. Chem. Soc. *2*, 980 (1949).
65. *Pascard-Billy, C.:* Acta Cryst. *18*, 829 (1965).
66. *Goypiron, A., De Villepin, J.:* Submitted for publication in Can. J. Spectr.
67. *Derissen, J. L.:* J. Mol. Struct. *7*, 67 (1971).
68. *Haurie, M., Novak, A.:* J. Chim. Phys. *62*, 146 (1965).
69. *Nitta, I., Tommiie, Y., Hoe Koo, C.:* Acta Cryst. *5*, 292 (1952).
70. *Lamrani-Karim, L.:* Thesis, University of Paris 1972.
71. *Novak, A., Saumagne, P., Bok, L. D. C.:* J. Chim. Phys. *60*, 1385 (1963).
72. *Almenningen, A., Bastiansen, O., Motzfeldt, T.:* Acta Chem. Scand. *23*, 2848 (1969).
73. *Millikan, R. C., Pitzer, K. S.:* J. Am. Chem. Soc. *80*, 3515 (1958).
74. *Holtzberg, F., Post, B., Fankuchen, I.:* Acta Cryst. *6*, 127 (1953).
75. *Tellgren, R., Olovsson, I.:* J. Chem. Phys. *54*, 127 (1971).
76. *Einspahr, H., Marsh, R. E., Donohue, J.:* Acta Cryst. *B 28*, 2194 (1972).
77. *Moore, F. H., Power, L. F.:* Inorg. Nucl. Chem. Letters *7*, 873 (1971).
78. *Nahringbauer, I.:* Acta Chem. Scand. *23*, 1653 (1969).
79. *Stanford, R. H.:* Acta Cryst. *23*, 825 (1967).
80. *Novak, A., Cotrait, M., Joussot-Dubien, J.:* Bull. Soc. Chim. *1965*, 2536.
81. *Skinner, J. M., Stewart, G. M. D., Speakman, J. C.:* J. Chem. Soc. *1954*, 180.
82. *Hadži, D., Novak, A.:* Infrared Spectra of and Hydrogen bonding in Some Acid Salts of Carboxylic Acids. University of Ljubljana (1960).
83. *Brown, C. J., Peiser, H. S., Turner-Jones, A.:* Acta Cryst. *2*, 167 (1949).
84. *Bacon,* G. E., *Curry, N. A.:* Acta Cryst. *9*, 82 (1956).
85. *Hamilton, W. C., Ibers, J. A.:* Acta Cryst. *16*, 1209 (1963).
86. *Snyder, R. G., Ibers, J. A.:* J. Chem. Phys. *36*, 1356 (1962).
87. *Thomas, J. O.:* Acta Cryst. *B 28*, 2037 (1972).
88. *De Villepin, J., Novak, A.:* to be published.
89. *Cotrait, M.:* Acta Cryst. *B 28*, 781 (1972).
90. *Novak, A., Cotrait, M., Joussot-Dubien, J., Lascombe, J.:* Bull. Soc. Chim. 1965, 1440.
91. *Ahmed, N. A. K., Liminga, R., Olovsson, I.:* Acta Chim. Scand. *22*, 88 (1968). *Nilsson, A., Liminga, R., Olovsson, I.:* Acta Chim. Scand. *22*, 719 (1968).
92. *Lindgren, J., De Villepin, J., Novak, A.:* Chem. Phys. Letters *3*, 84 (1969).
93. *Speakman, J. C.:* Structure and Bonding (ed. *J. D. Dunitz*), Vol. 12, p. 141. Berlin–Heidelberg–New York: Springer 1972.
94. *McAdam, A., Currie, M., Speakman, J. C.:* J. Chem. Soc. A *1971*, 1994.
95. *Hadži, D., Orel, B., Novak, A.:* Spectrochim. Acta *29A*, 1745 (1973).
96. *Manojlović, Lj., Speakman, J. C.:* Acta Cryst. *B 64*, 323 (1968).
97. *Grenié, Y.:* Thesis, University of Bordeaux, 1972.
98. *Haurie, M., Novak, A.:* J. Chim. Phys. *64*, 679 (1967).
99. *Collin, R. L., Lipscomb, W. N.:* Acta Cryst. *4*, 10 (1951).
100. *Durig, J. R., Bush, S. F., Mercer, E. E.:* J. Chem. Phys. *44*, 4238 (1966).
101. *Hall, D., Llewellyn, F. J.:* Acta Cryst. *9*, 108 (1956).
102. *Orville-Thomas, W. J., Parsons, A. E.:* Trans. Faraday Soc. *54*, 460 (1958).
103. *Stewart, R. F., Jensen, L. H.:* J. Chem. Phys. *40*, 2071 (1964).
104. *Kyogoku, Y., Higuchi, S., Tsuboi, M.:* Spectrochim. Acta *23A*, 969 (1967).
105. *Sakurai, K., Tomiie, Y.:* Acta Cryst. *5*, 293 (1952).
106. *Hoogsteen, K.:* Acta Cryst. *16*, 907 (1963).
107. *Martinez-Carrera, S.:* Acta Crysr. *20*, 783 (1966).
108. *Bellocq, A. M., Perchard, C., Novak, A., Josien, M. L.:* J. Chim. Phys. *62*, 1334 (1965).
109. *Watson, D. G., Sweet, R. M., Marsh, R. E.:* Acta Cryst. *19*, 573 (1965).

110. *Lautié, A., Novak, A.:* J. Chim. Phys. *65*, 1359 (1968).
111. *Speakman, J. C.:* J. Chem. Soc. *1949*, 3357.
112. *Jones, R. E., Templeton, D. H.:* Acta Cryst. *11*, 484 (1958).
113. *Karle, J., Brockway, L. O.:* J. Am. Chem. Soc. *66*, 574 (1944).
114. *Overend, J.:* Infrared Spectroscopy and Molecular Structure (ed. *M. Davies*), p. 351. New York: Elsevier Publishing Co. 1963.
115. *Thompson, W. E., Pimentel, G. C.:* Z. Elektrochem. *64*, 748 (1960).
116. *Lucazeau, G., Novak, A.:* J. Raman Spectry., *1*, 573 (1973).
117. *Bratos, S.:* private communication.
118. *Jones, D., Evans, D. F.:* Nature *199*, 277 (1963).
119. *Hadži, D.:* private communication.
120. *Okaya, Y., Pepinski, R.:* Acta Cryst. *10*, 324 (1957).
121. *Shrivastava, H. N., Speakman, J. C.:* J. Chem. Soc. *1961*, 1151.
122. *Hamilton, W. C.:* Acta Cryst. *15*, 353 (1962).
123. Tables of Interatomic Distances and Configuration in Molecules and Ions. Supplement 1956—1959 (ed. *L. E. Sutton*) Special publication n° 18. London: Chemical Society 1965.
124. *Herzberg, G.:* Molecular Spectra and Molecular Structure. I. Spectra of Diatomic Molecules. Princeton, New Jersey: D. Van Nostrand Company Inc. 1964.
125. — Molecular Spectra and Molecular Structure. II. Infrared and Raman Spectra of Polyatomic Molecules. Princeton, New Jersey: D. Van Nostrand Company Inc. 1964.
126. *Lascombe, J., Devaure, J., Josien, M. L.:* J. Chim. Phys. *61*, 1271 (1964).
127. *Huong, P. V., Turrell, G.:* J. Mol. Spectry. *25*, 185 (1968).
128. *Iogansen, A. V., Kurkchi, G. A.:* Opt. Spektrosk. *13*, 480 (1962).
129. *Perchard, J. P., Josien, M. L.:* J. Chim. Phys. *65*, 1856 (1968).
130. *Le Calvé, N., Garrigou-Lagrange, C.:* J. Chim. Phys. *65*, 1017 (1968).
131. *Perchard, C., Bellocq, A. M., Novak, A.:* J. Chim. Phys. *62*, 1344 (1965).
132. *Lautié, A., Novak, A.:* Can. Specty. *17*, 113 (1972).
133. *Lake, R. F., Thompson, H. W.:* Proc. Roy. Soc. (London) *291*, 469 (1966).
134. *Hadži, D., Novak, A., Gordon, J. E.:* J. Phys. Chem. *67*, 1118 (1963).
135. *Nyquist, R. A.:* Spectrochim. Acta *19*, 1655 (1963).
136. *Perchard, C., Novak, A.:* J. Chim. Phys. *65*, 1964 (1968).
137. *Foglizzo, R., Novak, A.:* J. Chim. Phys. *66*, 1539 (1969).
138. — — Appl. Spectry. *24*, 601 (1970).
139. *Ginn, S. G. W., Wood, J. L.:* Spectrochim. Acta *23A*, 611 (1967).
140. *Foglizzo, R., Novak, A.:* J. Mol. Struct. *7*, 205 (1971).
141. *Stanevich, A. E.:* Opt. Spectry. (USSR) (English Transl.) *16*, 243 (1964).
142. *Carlson, G. L., Witkowski, R. E., Fateley, W. G.:* Spectrochim. Acta *22*, 1117 (1966).
143. *Speakman, J. C.:* M. T. P. International Review of Science Physical Chemistry, Series one, *11*, Chemical Crystallography, p. 1. London: Butterworth and Co., Ltd. 1972.
144. *Hadži, D., Novak, A.:* Nuovo Cimento 2 (X), Supplement *1955*, 715.
145. *Le Narvor, A., Saumagne, P., Novak, A.:* J. Chim. Phys. *64*, 1643 (1967).
146. *Singh, T. R., Wood, J. L.:* J. Chem. Phys. *50*, 3572 (1969).
147. *Savelev, V. A., Sokolov, N. D.:* Teor. Eksperim Khim. Akad. Nauk Ukr. SSR *6*, 174 (1970).
148. *Rundle, R. E.:* Physique *25*, 487 (1964).
149. *Costain, C. C., Srivastava, G. P.:* J. Chem. Phys. *41*, 1620 (1964).

150. *MacDonald, A. L., Speakman, J. C., Hadži, D.:* J. Chem. Soc. Perkin II *1972*, 825.
151. *Sheppard, N.:* Comment at the International Conference on Hydrogen Bonding. Ottawa 1972.
152. *Ibers, J. A.:* J. Chem. Phys. *41*, 25 (1964).
153. *Stepišnik, J., Hadži, D.:* J. Mol. Struct. *13*, 307 (1972).
154. *Perchard, C.:* Thesis, University of Paris 1968.
155. *Perrier, A., Julien-Laferrière, S., Lebas, J. M.:* J. Chim. Phys., *70*, 1684 (1973).
156. *Thomas, J. O.:* Acta Cryst. *B 29*, 1767 (1973).

Received December 6, 1973

Structure and Bonding: Index Volume 1-18

GPSR Compliance

The European Union's (EU) General Product Safety Regulation (GPSR) is a set of rules that requires consumer products to be safe and our obligations to ensure this.

If you have any concerns about our products, you can contact us on ProductSafety@springernature.com

In case Publisher is established outside the EU, the EU authorized representative is:

Springer Nature Customer Service Center GmbH
Europaplatz 3
69115 Heidelberg, Germany

Batch number: 10620141

Printed by Printforce, the Netherlands